THE
BABYLONIAN
THEOREM

THE
BABYLONIAN
THEOREM

THE MATHEMATICAL JOURNEY
TO PYTHAGORAS
AND EUCLID

PETER S. RUDMAN

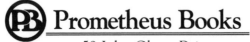
Prometheus Books

59 John Glenn Drive
Amherst, New York 14228-2119

Published 2010 by Prometheus Books

Inquiries should be addressed to
Prometheus Books
59 John Glenn Drive
Amherst, New York 14228–2119
VOICE: 716–691–0133
FAX: 716–691–0137
WWW.PROMETHEUSBOOKS.COM

14 13 12 11 10   5 4 3 2 1

Library of Congress Cataloging-in-Publication Data

Rudman, Peter Strom.
    The Babylonian theorem : the mathematical journey to Pythagoras and Euclid / by Peter
S. Rudman.
        p. cm.
    Sequel to: How mathematics happened, the first 50,000 years. 2007.
    ISBN 978-1-59102-773-7 (cloth : alk. paper)
    1. Mathematics, Babylonian. 2. Mathematics, Ancient. 3. Mathematics—Egypt—
History. 4. Pythagoras. 5. Euclid. I. Rudman, Peter Strom. How mathematics happened, the
first 50,000 years. II. Title.

QA22.R859 2010
510.935—dc22

                                                                                            2009039196

Printed in the United States of America

CONTENTS

LIST OF FIGURES

MATHEMATICAL SYMBOLS

$=$	equal to
$\cong$	approximately equal to
$\geq$	greater than or equal to
$\leq$	less than or equal to
$\propto$	proportional to
$\Rightarrow$	implies, leads to
$\sum$	summation, $\sum_{n=1}^{n=3} n = 1 + 2 + 3$
Δx	change in x, usually a small change
ε	a small quantity
$\int$	integration, summation (calculus notation)
dx	an infinitely small change (calculus notation)
QED	*quod erat demonstradum*, denotes end of a proof

GREEK NOTATION

A, B, C ... locations on diagrams

ABC   a triangle

ABC   an angle with location B at the apex

AB, *CD*   . . . line segments

PREFACE

This book is a sequel to *How Mathematics Happened: The First 50,000 Years*, which addressed the evolution of number systems and arithmetic. That history ended with some OB problem text calculations (OB = Old Babylonian, ca. 2000 to 1600 BCE). An in-depth look at these calculations is the starting point for this book.

Since I use the word *algebra* frequently, in order to avoid semantic quibbling, let me define my usage, which is conventional, with some simple examples:

- When I write $35 = 7 \times 5$, that is *arithmetic*.
- When I write "something is the product of the first quantity and the second quantity," that is *rhetorical algebra*.
- When I write, referring to a diagram of a rectangle, "area is the product of 7 and 5," where the length is given as 7 and the width is given as 5, that is *geometric algebra*, oftentimes referred to as *metric geometric algebra*.
- When I write, referring to a diagram of a rectangle, "area is the product of length and width," that is *rhetorical geometric algebra*.
- When I write "$A = a \times b$," that is *elementary symbolic algebra*, or simply *algebra*.
- When I write "$A = a \times b$," where the symbols are defined by a diagram of a rectangle, that is also *geometric algebra*, oftentimes referred to as *nonmetric geometric algebra*.

With these definitions at hand, I can succinctly summarize the contents of this book as tracing the evolution of mathematics from the *metric geometric algebra* of Egypt and Babylon ca. 2000 BCE to the *nonmetric geometric algebra* of Euclid ca. 300 BCE.

The number system affects how arithmetic is done; how arithmetic is done affects subsequent mathematical development. *How Mathematics Hap-*

pened explained why Egyptian and Babylonian number systems and arithmetic evolved differently so that it is now possible to understand the differences in their subsequent mathematics. For the benefit of those who have not read *How Mathematics Happened*, I start this book with three chapters that briefly summarize these previously developed essential concepts.

By the end of the nineteenth century, translations of documents excavated from Egypt and Babylon showed that by about by 2000 BCE the practical arithmetic and geometry required for the constructions and the administration of these empires had been learned—to nobody's great surprise. But in the 1930s, Otto Neugebauer (1899–1990) translated hitherto unintelligible OB cuneiform *problem texts* that unambiguously showed that OB scribes were using the Pythagorean theorem—some 1,500 years before its credited discovery by Pythagoras. That was a surprise. He also concluded that the OB scribes had solved quadratic algebra problems that had been considered a much later invention, which was an even greater surprise.

My opinion, first expressed in *How Mathematics Happened*, is that these "quadratic algebra" OB problem texts are not *algebra* but rather *geometric algebra*, derived from visualizations of two-dimensional geometric diagrams. My opinion did not derive from new translations that also came to this same conclusion, but rather because visualization is simply a natural and intuitive way of doing mathematics (and physics and chemistry). However, such was the stature of Neugebauer that his conclusion that OB scribes had done quadratic algebra was a largely unchallenged canon for most of the twentieth century. To make my point against such a formidable authority, I carefully analyze a selection of OB-era problem texts from both Egypt and Babylon and show that geometric visualization was the basis for essentially all of their mathematics, including such problems as arithmetic sequences, geometric sequences, and square roots that are usually thought of as number-theory problems. My analyses also illustrate the mathematical par and intellectual interaction between Egypt and Babylon because I show that both used the same algorithms. I have been unable to establish who copied from whom.

My analyses of OB "quadratic algebra" problem texts also provide the answer to the question of the origin of OB awareness of the Pythagorean theorem. I have interpreted a basic OB "quadratic algebra" visualization as what I call the **Babylonian theorem**:

For any right triangle (*a*, *b*, *c*) it is possible to construct another right triangle with sides: $\sqrt{4ab}$, $a - b$, $a + b$.

Figure P.1 illustrates application of this theorem to a right triangle with sides $a = 4$, $b = 1$ to produce a (3, 4, 5) Pythagorean triple. I show how the Babylonian theorem fathered OB awareness of the existence of the Pythagorean theorem, the most famous and useful of all equations in mathematics. If Pythagoras gets recognition for a theorem that he did not discover, certainly the biological father of the Pythagorean theorem deserves the recognition I have given here.

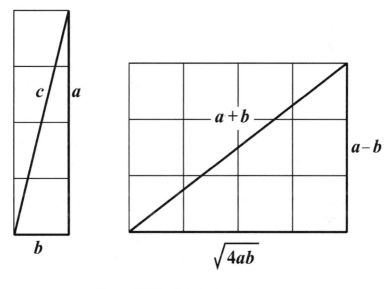

Figure P.1 The Babylonian theorem

OB-era problem texts were generally either a teacher's notes to himself or a student's notes to himself or a student's solutions of problems set by the teacher. The documents were numerical calculations not intended for understanding by others; OB-era communication of mathematics required person-to-person contact. This was not an efficient system for expanding knowledge and OB mathematics stagnated by about 1600 BCE. There was then a mysterious hiatus in recovery of mathematical documentation until about 500 BCE, when very similar but somewhat more sophisticated documentation reappeared, known as LB (LB = Late Babylonian). This was an age of increased contact with Greek civilization and it was the slightly improved OB metric geometric algebra that was inherited by the Greeks.

The Greeks did not squander their inheritance. Already in the sixth century BCE, Pythagoras digitalized his inheritance of LB geometric algebra and invented number theory. I show how he produced the same expression for Pythagorean triples that OB scribes had obtained and that this was probably the origin of Pythagoras's awareness of the Pythagorean theorem, rather than the Euclid-like rigorous proof that has been baselessly credited to him.

It is difficult to identify LB influence in Greek mathematics because geometric visualization was just as natural and intuitive for the Greeks as it was for the Babylonians and Egyptians. Thus, even though an OB-era scribe used the circumference of a hexagon to approximate the circumference of a circle and a millennium later Archimedes used exactly the same concept, my guess is that was just the same intuitive independent visualization. I conclude that what enabled Archimedes to do a much better calculation than his OB predecessor and in general enabled the Greeks to far surpass the nearly stagnant OB mathematics was simply the insertion of alphabetic notation onto the visualized geometric figures. The importance of this Greek invention has not been appreciated because the major translations and analyses of Greek mathematics were made before translations of OB-era problem texts became available and thus the difference alphabetic notation made was not obvious. It was as if geometry had been born with alphabetic notation and it was not conceived how it could have been done without it.

Such symbolic notation was natural for users of an alphabetic language; the Egyptians and Babylonians, without alphabetic languages, never could have invented symbolic notation. Limited to defining their mathematics by metric geometric algebra, their mathematicians were also limited in their ability to communicate and in the complexity of problems they could consider.

To illustrate the universality of the intuitive visualizations and the effect of alphabetic notation on mathematics, I have traced the evolution of some problems forward from OB-era origins. For example, I trace the evolution of determining square roots from Babylon and Egypt to India to Pythagoras to Euclid to Archimedes to Ptolemy. Incidentally, how Archimedes actually did his calculation has been somewhat of a mystery. I present a new algorithm that I think better explains his method.

This book is not an encyclopedic presentation of *what* mathematics was done. For the most part, the *what* is undisputed. I am interested in understanding *how* and *why* they did the math, which is seldom undisputed. One can marshal the facts and find revealing patterns and fit what one thinks are logical explanations, but in the end it is a guess that in mathematical jargon is called a *conjecture*. This book collects my conjectures about *how* and *why* math was done from two to four thousand years ago. Have I got it right? Now it's your turn to guess.

To make the read easier, I express all the derivations in one consistent algebraic notation and I do all numerical calculations in modern base-10. I have simplified the scholarly literature to a more digestible level without sacrificing rigor. I never resort to "it can be proved" but give algebraic derivations of all calculations in sufficient detail so that someone who wants to completely understand the *how* can. For those who find wading through a sequence of algebraic equations boring, they can be skipped and the text will painlessly give the conclusions.

Moderate concentration and standard high school algebra and geometry are sufficient for understanding this book. For those who enjoy recreational math problems and wish to deepen and test mastery of new concepts, distributed throughout the book are FUN QUESTIONS. Appendix A gives the answers.

I hope that you find this book provides useful insights into how mathematics evolved and it will make your pursuit of mathematics more enjoyable.

ACKNOWLEDGMENTS

This book is primarily about the evolution of mathematics from about the twentieth century BCE through the first century CE. It is also about the evolution of modern interpretations of such mathematics.

Much of the translations and interpretations of the cuneiform tablets are the remarkable work of one man, Otto Neugebauer, during the 1930s. However, in this book I think that I convincingly show how much he overestimated Babylonian algebraic prowess. What makes the history of the modern interpretation of Babylonian mathematics so interesting is how a mathematician of Neugebauer's stature could have made this mistake, while in so many other cases his analyses and intuition were brilliant, and why his algebraic interpretations were significantly unchallenged until near the end of the twentieth century.

Unfortunately, Otto Neugebauer (1899–1990) and his students Bartel van der Waerden (1903–1996) and Asger Aaboe (1922–2007), who might have been able to unambiguously resolve these questions, had all died before I was able to question them, but both Waerden and Aaboe have also written books about Babylonian mathematics so I have had to make do with what I could glean from them. However, I have solicited and received comments from several noted scholars. Some of these comments I have referred to in the text and here I wish to thank all who have responded to my queries regarding Neugebauer or to questions I had about their work.

Noel Swerdlow, who worked with Neugebauer for some twenty years starting in the 1960s, is the only one I was able to find who knew Neugebauer personally. According to Swerdlow, their work together was on ancient astronomy and they did not discuss Babylonian mathematics, but he provided useful insights about him. Scholars Joran Friberg, Jens Hoyrup, and Eleanor Robson have all provided valued information and opinions, but all conclusions herein are mine and I absolve everybody of any errors of commission or omission.

Chapter 1

NUMBER SYSTEM BASICS[1]

W e can be sure that the place-value, base-10 number system with Hindu-Arabic symbols that we now use globally must be just about the best for everyday use. It is the clear winner of a survival-of-the-fittest number-system game that has been going on for millennia. (The nomenclature *base-10* is interchangeable with *decimal* and *place-value* with *positional*.) The victory of base-10 is remarkable because the choice was simply because we happen to have ten fingers. A creationist might see a divine hand literally at work here, but the rest of us require another explanation. What makes base-10 such an ideal number system?

The larger the base, the more compact the number system. For example, the base-10 number 189 (= $1 \times 10^2 + 8 \times 10 + 9$) is written as 10111101 (= $2^7 + 2^5 + 2^4 + 2^3 + 2^2 + 1$) as a base-2 number and as 99 (= $9 \times 20 + 9$) as a base-20 number. Base-10 may not be as compact as base-20 but it is adequately compact for everyday use. Electronic digital computers use base-2 because the two symbols 0 and 1 conveniently represent OFF and ON switches. Base-20 evolved in cultures that first solved the problem of counting to greater than ten by adding toes.

Any N-position, base-b integer can be expressed using Hindu-Arabic symbols in algebraic notation as a sequence of N integers: $a_{N-1}a_{N-2} \dots a_n \dots a_2a_1a_0$, where the a_n's are integers from 0 to $b - 1$. Equation (1.1) gives the base-10 value of an integer in any base.

$$a_{N-1}b^{N-1} + a_{N-2}b^{N-2} \dots + a_n b^n + \dots + a_2 b^2 + a_1 b + a_0 \qquad (1.1)$$

The fractional part of any N-position, base-b number can be expressed using Hindu-Arabic symbols in algebraic notation as a sequence of N integers: $0.a_{-1}a_{-2} \dots a_{-N}$. Equation (1.2) gives the base-10 value of the fractional part of a number in any base:

$$a_{-1}b^{-1} + a_{-2}b^{-2} \dots + a_{-N}b^{-N} \qquad (1.2)$$

FUN QUESTION 1.1: Write the base-5 number 234 as a base-10 number.

FUN QUESTION 1.2: Write the base-10 number 189 as a base-5 number. Hint: $189 = 125a_3 + 25a_2 + 5a_1 + a_0$. Start by finding the largest possible value for a_3.

FUN QUESTION 1.3: Write the base-5 fractional number 0.234 as a base-10 number.

FUN QUESTION 1.4: Write the base-10 fractional number 0.189 as a base-5 number.

For the Babylonian base-60 system, various methods enable expression with just the ten Hindu-Arabic symbols. The system I shall use is the same notation generally used to express time (hours:minutes:seconds with 1 hour = 60 minutes = 60^2 seconds), a vestige of the Babylonian base-60 system. Colons separate place values and a period (called a *sexagesimal point*) separates the integer part from the fractional part (just as a *decimal point* does in the base-10 system). Thus, the *decimally transcribed* sexagesimal (base-60) number $1:3:20.15:40_{60}$ has a base-10 value: $1 \times 60^2 + 3 \times 60 + 20 + 15/60 + 45/60^2 = 3800.2625$. Lacking any generally agreed-upon convention, there is some unavoidable confusion in reading decimally transcribed base-60 numbers. In this book, all calculations presented are done in modern base-10 to avoid the distraction of having to interpret other notations. Whenever there is the possibility of confusion as to whether a number is written in decimally transcribed sexagesimal, I add a subscripted 60 as just previously done.

The larger the base, the larger the memorization burden it imposes. The memorization of the values of only ten symbols (0, 1, 2, 3, 4, 5, 6, 7, 8, 9) in the base-10 system is no problem and neither would be the memorization of the b symbols in any practical base-b number. However, when arithmetic, and particularly multiplication, is considered, the memorization burden of a number system with a large base looms large. Figure 1.1 presents the base-10 multiplication table that we all boringly learned in elementary school but now barely ever use since electronic calculators became ubiquitous. We do not have to memorize all 102 possible products because most of the entries are eliminated by the rules $0 \times n = 0$, $1 \times n = n$, and $m \times n = n \times m$, where m and n are any numbers. Without much effort, we can simply add up the entries in figure 1.1 and find that there are 36, not an overwhelming memorization burden.

x	2	3	4	5	6	7	8	9
2	4	6	8	10	12	14	16	18
3		9	12	15	18	21	24	27
4			16	20	24	28	32	36
5				25	30	35	40	45
6					36	42	48	54
7						49	56	63
8							64	72
9								81

Figure 1.1 Multiplication table—decimal system

In order to calculate the number of multiplication table entries for any base-b, we note from figure 1.1 that the number of entries is the arithmetic series $1 + 2 + \ldots + 7 + 8 = 36$, and by generalizing this series we can obtain that for base-b, the number of multiplication table entries is

$$N_b = \frac{(b-2)(b-1)}{2} \tag{1.3}$$

Using this equation, we can readily calculate that for base-20, $N_{20} = 171$ is quite a memorization burden, while for the Babylonian base-60, $N_{60} = 1,711$ imposes an impossible memorization burden, which explains why these number systems did not survive.

Now we can see why base-10 is essentially an optimized number system: the base is large enough to produce adequately compact numbers, but small enough not to impose too heavy a memorization burden. For the Babylonians, multiplication with a base-60 number system was indeed a problem and we shall consider later how they solved it, but now we want to understand why they invented a number system with such a large base. The answer is simply that they invented their number system in a prehistoric (prewriting), prearithmetic era. For counting, base-60 was fine, but when their culture advanced to a stage where arithmetic was required, base-60 was so entrenched in Babylonian habits and records that they preferred to find ways to cope rather than change their number system. The problem is similar to current reluctance in the United States to abandon use of archaic English units and adopt the metric system.

Let us go back to the time of the so-called Neolithic revolution, around 10,000 BCE in Egypt and Babylon, when hunter-gatherer cultures were beginning to transform into materially richer and more complex herder-farmer cultures. Finger counting now no longer satisfied counting requirements, and the ancient solution to its limitations was pebble counting. In its simplest realization, pebble counting requires little memorization and uses only the principle of one-for-one correspondence. Consider a shepherd putting a pebble in a bowl as each sheep goes out to pasture and removing a pebble as each sheep returns. The number of pebbles left in the bowl is then the number of sheep lost. This is counting and arithmetic without a need for names for numbers, or an ability to articulate counting, with no limit to the number counted, and with a permanent record, a significant advance beyond finger counting.

With growing understanding of the concept of number, our generic shepherd (now perhaps many generations later) wants to know how many sheep he has. A bowl of hundreds of pebbles is not very defining, so he replaces each group of ten pebbles with one larger pebble. Why ten? Because he had first learned to count using his fingers, just as children still do, and therefore ten is a natural choice. If he still has too many pebbles to define his number of sheep conveniently, he can replace every ten big pebbles with a bigger pebble, and so on. There are now never more than nine pebbles of the same size, so even a quick glance at his collection of pebbles allows him to visualize the number. Unaware though he surely was, he had now invented an *additive*, base-10 number system. The smallest pebble has a value of one; the next larger pebble has a value of ten; the next larger pebble after that has a value of a hundred; and so on. A sequence of replacements using the same replacement number defines a number system with a *base* equal to the replacement number. Figure 1.2 illustrates this unrealized and unintended pebble-counting invention of an *additive, base-10 number system*.

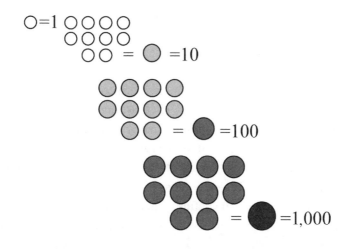

Figure 1.2 Pebble counting with 1-for-10 replacements

If rather than using pebbles of different sizes in 1-for-10 replacements, ten pebbles in one bowl were replaced by one pebble in a different bowl, a *place-value, base*-10 *number system* would have been invented. Even if one starts with an additive system with different-sized pebbles, for easier visualization of quantity one naturally and intuitively tends to gather each size in a separate pile. Now the values of pebbles are redundantly defined both by pebble size and by pebble position. Eventually this unnecessary redundancy tends to be realized, and hence additive systems tend to evolve into *place-value* systems.

If his hands were man's first calculating machine, then pebbles were his second. Number systems with a base originated in a pebble-counting era. Once a replacement is chosen for some logical reason, such as 1-for-10 because we have ten fingers, then why change it in a sequence of replacement numbers. Since a sequence of the same replacements defines a number system with a base, number systems with a base tend to evolve naturally once use of replacement has begun. The advantage of a number system with a base becomes more apparent when arithmetic other than simple addition of small numbers is required.

We have now developed the three important characteristics of an efficient number system, and they are valid whether the counting is by fingers, pebbles, or written symbols:

1. Replacement—extends the counting limit.
2. Base—enables efficient arithmetic.
3. Position—minimizes the number of different symbols.

Quantitative measurement, in addition to just counting of things, became a requirement after about 10,000 BCE. This imposed new demands on number systems. Consider the following scenario: a woman is making a garment and invents the natural, intuitive measuring system of using the widths of her fingers. She measures something as 19 finger widths. In practice, measurement in units of finger widths is by a series of handbreadth-to-handbreadth placements, so she defines a handbreadth as a natural, intuitive, new unit, with 1-for-4 replacements of fingers by hands. Now, letting a larger pebble represent a handbreadth, she records her measurement as 4 large pebbles and 3 small pebbles. Our prehistoric seamstress has now invented a measurement system with units of handbreadths and fingers. The 1-for-10 replacement scheme that was so natural and intuitive for counting is no longer obviously the better system for units of measurement. What to do?

• Solution 1: Retain the natural 1-for-10 replacements for counting, and retain the various natural replacements for measuring. This method largely accounts for the English system still used in the United States although not officially in England since 1965.

• Solution 2: Define new, not-so-natural units of measurement so that 1-for-10 replacements also define measurement units. This is an important component of the present, almost globally adopted metric system and was also the ancient Egyptian solution. It might be an overstatement to say that the Egyptians realized more than 5,000 years ago what the rest of the world is only now realizing. It is possible they simply did what came naturally and used the same intuitive 1-for-10 replacement system for measuring units as they did for counting. It was fortuitously a good choice.

• Solution 3: Modify the natural 1-for-10 replacements for counting by adopting the natural replacements for measurements. This was the Babylonian solution. It was fortuitously a bad choice that would only become evident millennia later when required to perform arithmetic that was more complex.

units	inch (thumb)		hand		foot		yard (arm)		fathom		rod
							body parts		extended body parts		
replacement numbers		4		3		3		2			
		12					15				

Figure 1.3 English body-parts length units (fifteenth century)

Units of linear measurement started just as simple counting did, by using body parts. Using the example of English units, with which we are most familiar, figure 1.3 shows the replacement numbers as they were defined up to the fifteenth century. To convert from a larger unit to a smaller unit it is necessary to multiply by the product of all the replacement numbers between the respective units. Thus, from figure 1.3 we can calculate that the number of inches in a fathom is $4 \times 3 \times 3 \times 2 = 72$. Compared to converting units within the modern metric system where conversion is only a matter of moving a decimal point (for example, 1.75 m = 175 cm), English units are a bother. Nowadays, with the ready availability of rulers and tape measures there is no logical reason to continue to use body-parts units, but conversion is a question of politics, not mathematics, and hence is beyond the scope of this book.

For longer distances, the fundamental English unit was based on farming practice, the *furlong*, literally a furrow long. Another example of an English unit based on farming practice is the *acre*, which legend has it was the area plowed by a pair of oxen in one day. Eventually, as measurements that were more precise were required, a government had to set standards to reconcile farming-practice units and body-parts units. Thus, up to the fifteenth century in England the furlong was defined as 1 furlong = 40 rods = 600 feet. However, England had also inherited from the Romans the unit of the mile, which the Romans had defined as 1,000 double paces. A double pace is 5 feet, so the mile, clearly derived from the Latin word for 1,000, was 5,000 feet. This had the unfortunate affect of inconveniently making the mile a nonintegral (5,000/600 = 8.333...) multiple of the rod, so in the sixteenth century Queen Elizabeth redefined many units to essentially their present values, which among other things redefined the mile as exactly 8 furlongs. The details of this redefinition need not concern us here. The essential point is that every culture goes through a process similar to what occurred centuries ago in England. Units of measurement from different sources must be reconciled by standards set by the government.

For volume measurements, for which body parts do not provide convenient units, ancient practice was that some frequently used container size was chosen as a reference and other units were defined as multiples of this. What is interesting about this method is the universal tendency to choose multiples of two. Figure 1.4 illustrates such practice with English volume units.

1 gallon				128 ounces	
1/2 "	2 quarts			64 "	
1/4 "	1 "			32 "	
1/8 "	1/2 "	1 pint		16 "	
1/16 "		1/2 "	1 cup	8 "	
1/32 "			1/2 "	4 "	
1/64 "			1/4 "	2 "	
1/128 "				1 "	16 drams
1/256 "					8 "
1/512 "					4 "
1/1,024 "					2 "
1/2,048 "					1 "

Figure 1.4 US liquid volume standards—English units

In ancient Egypt the basic body-parts unit was the cubit, the distance from the elbow to the tip of the middle finger, about 46 cm. Figure 1.5 illustrates its subdivision into fingers, handbreadths, and feet.

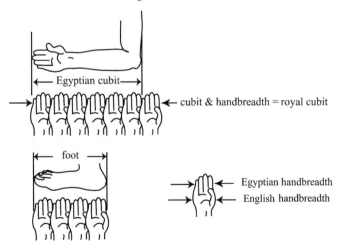

Figure 1.5 Egyptian body-parts units

Figure 1.5 also illustrates the peculiar definition of a *royal cubit* that is equal to 7 handbreadths, whereas the cubit is equal to 6 handbreadths. No surviving document explains the need for the royal cubit unit. We can only guess what the reason was. Perhaps the unit was to differentiate between properties of the pharaoh and properties of common people. However, a clue is provided by the definition of another anomalous unit, the remen: 1 remen = 5 handbreadths. My guess is that the role of these apparently anomalous units was to enable easy halving or doubling of land areas. If the linear dimensions of an area were first measured as a certain number of royal cubits, a half-size area would be one with its linear dimensions given as the same number of remens. If the linear dimensions of an area were first measured as a certain number of remens, a doubled area would be one with its linear dimensions given as the same number of royal cubits. Figure 1.6 illustrates the process for various shapes of areas: proceeding from left to right the area is reduced by $(5/7)^2 = 25/49 \cong 1/2$; proceeding from right to left the area is increased by $(7/5)^2 = 49/25 \cong 2$.

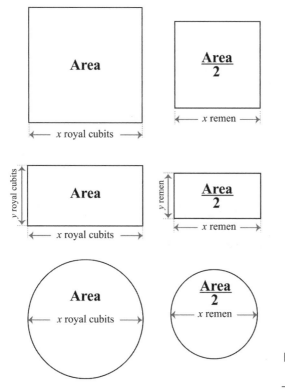

Figure 1.6 Egyptian halving and doubling of areas

For longer distances and larger areas, the Egyptians mimicked their base-10 counting system as illustrated in figure 1.7. The tA is both a distance and an area unit since it is a line with a finite width. Such use of a unit for both distance and area was general ancient practice.

Volume units for liquids were also base-10: hin $\cong$ 0.5 liter, hqAt $\cong$ 5 liter, Xar $\cong$ 50 liter. However, volume units for grain were based on successive halving. Starting with a hqAt, fractional volumes were called Horus-eye fractions and were written with special symbols as illustrated in figure 1.8. It was documented practice in Babylon and Canaan to use the same unit for a land area and the volume of grain required to seed the area. Perhaps the system of halving land area as illustrated in figure 1.6 and the system of halving grain volumes exhibited by the Horus-eye fractions of figure 1.8 are related.

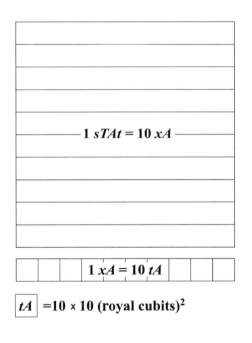

Figure 1.7 Land-area units in ancient Egypt

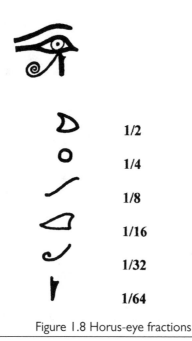

🜄	**1/2**
⊙	**1/4**
╱	**1/8**
◁	**1/16**
◡	**1/32**
❘	**1/64**

Figure 1.8 Horus-eye fractions

In Babylon, the basic body-parts unit was also the cubit, generally taken to be about 50 cm, about midway between the Egyptian values for the cubit and the royal cubit. Definition of land area was certainly one of the earliest important uses of units of measurement, and the basic Babylonian length unit for defining long distances was the extended-body-parts unit, the nindan: 1 nindan = 12 cubits. Figure 1.9 illustrates the various length/area units used in Babylon. All units are defined as multiples of the basic unit the sar, 1 sar = 1 nindan². In conformity with intuitive counting practice, the next larger unit is the eshe, 1 eshe = 10 sar. However, in order to reconcile the eshe with the length/area farming-practice unit of the USH, the USH was redefined as an exact multiple, 1 USH = 6 eshe, and then this practice of a sequence of alternating multiples of 10 and 6 was continued.

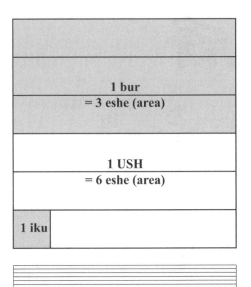

1 eshe (area) = 10 USH (length & area)

1 USH (length & area) = 6 eshe (length & area)

1 eshe (length & area) = 10 sar

1 square nindan = 1 sar

Figure 1.9 Land-area units in Babylon

The Babylonian definition of land-measurement units is comparable to the process previously considered of defining English units. For ready comparison, figure 1.10 presents this reconciliation of the English extended-body-parts unit, the rod, with the farming-practice units of the furlong and the acre. The fact that a square furlong is defined as exactly 10 acres probably means that the unit of the acre was a redefinition of a more ancient unit based on the amount of land that could be plowed in a day by a team of oxen. It may at first glance appear weird to be comparing English units in the sixteenth century with Babylonian units of some 5,000 years earlier, but the common denominator in farming practice throughout this long interval was plowing with oxen.

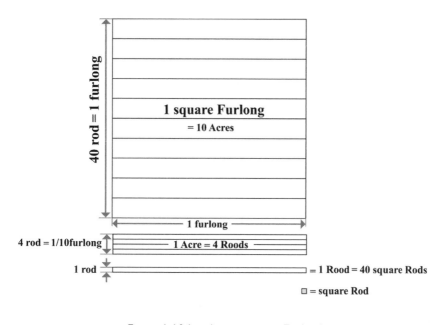

Figure 1.10 Land-area units in England

This system of measuring units with an alternating sequence of 1-for-10 and 1-for-6 replacements was also adopted for the Babylonian number system, producing essentially a metric system with easy recording and calculation of length, area, and volume. Eventually it was realized that a system of alternating 1-for-10 and 1-for-6 replacement numbers was equivalent to the base-60 number system and it was treated as such. Thus, to the question of why a number system with such a ridiculously large base was chosen, the answer is that it was not chosen. Rather, what appeared to be an advantageous small change from a sequence of just 1-for-10 replacements to a sequence of alternating 1-for-10 and 1-for-6 replacements was chosen. When Babylonian numbers are decimally transcribed as base-60 numbers, it obscures the fact that the number system is really an alternating sequence of 1-for-10 and 1-for-6 replacements and this has misled many who have sought explanations for the choice of base-60.

Chapter 2

EGYPTIAN NUMBERS AND ARITHMETIC[1]

T he fourth millennium BCE was a period of great change in Egypt and Mesopotamia: oxen were first hitched to a plow, extensive use of irrigation began, ceramic technology matured, and the use of copper and bronze tools began. Better farming productivity made it possible to feed an urban population of artisans, merchants, soldiers, priests, and of course tax collectors; cities began to grow. Now there was a need for more efficient record keeping, and the solution was writing, which began by around 3000 BCE in both Egypt and Mesopotamia.

The surviving mathematical record of ancient Egypt is very lean. Most of what is known has been obtained from just one document, the *Rhind Mathematical Papyrus* (referred to as *RMP* by the cognoscenti, and henceforth here). It is named after Henry Rhind, a collector of antiques who purchased it in 1858. A scribe named Ahmes wrote the papyrus in about 1700 BCE and so some experts refer to it as the Ahmes Papyrus. Ahmes wrote that he had copied it from a papyrus written about a hundred years earlier, but much of the content probably relates to understanding from even earlier times. It contains a table of fractions and eighty-seven mathematical problems, presumably a teacher-scribe's lecture notes.

Egyptian written numbers replicated pebble counting by simply substituting symbols for pebbles. Figure 2.1 exhibits some of the *hieroglyphic* number symbols. Reading hieroglyphic numbers is obvious. There were other symbols for higher powers of ten, so that any number, however large, could be written just by defining new symbols. There is no limit to the number of hieroglyphic symbols required, but use of numbers higher than one hundred thousand was rare, so in reality it was necessary to memorize values of only about six symbols. Hieroglyphic writing was generally used only for inscriptions on monuments and tombs. It frequently accompanied artwork, and since it was an additive system, the symbols could be arranged in any order, left to right or right to left, to fit the artistic content without confusing the value.

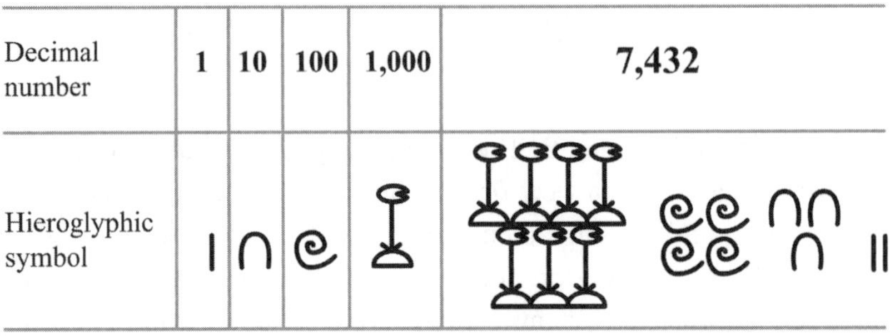

Decimal number	1	10	100	1,000	7,432
Hieroglyphic symbol					

Figure 2.1 Egyptian hieroglyphic number symbols

Administrative and literary documents were generally written on papyrus in cursive script called *hieratic*. *RMP* is written in hieratic. In addition to making writing faster than with hieroglyphic symbols, hieratic numbers obviated the need for repetitions of the same symbol. Figure 2.2 presents one variation of hieratic number symbols, which like cursive script in any language tends to vary with writer and over time. Note that writing numbers up to 9,999 in hieratic notation requires memorizing 36 symbols, not a particularly heavy memorization burden. Hieratic, like script in all Semitic languages, is always written from right to left and thus all numbers in hieratic script appear to be written backward to those of us who are used to writing derived from Latin.

When we do pencil/paper addition with our place-value, base-10 system we use the memorized addition table with the 45 entries exhibited in figure 2.3. If the Egyptians had done brush/papyrus addition with a hieratic, base-10 addition table, the table would have required 45 entries for each column of 9 symbols in figure 2.2. To add numbers up to 9,999 would thus have required $4 \times 45 = 180$ entries. This is now a bit of a memorization burden, and no such tables have ever been found, so memorized addition tables were probably not used.

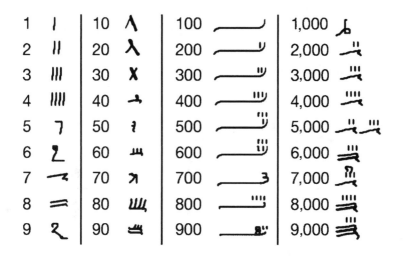

Figure 2.2 Hieratic number symbols (ca. 1200 BCE)

+	1	2	3	4	5	6	7	8	9
1	2	3	4	5	6	7	8	9	10
2		4	5	6	7	8	9	10	11
3			6	7	8	9	10	11	12
4				8	9	10	11	12	13
5					10	11	12	13	14
6						12	13	14	15
7							14	15	16
8								16	17
9									18

Figure 2.3 Addition table—decimal system

Presumably, the Egyptians added with an abacus something like that illustrated in figure 2.4, which is just a device that makes for easy implementation of the basic operation of additive, base-10 pebble counting: whenever the number of pebbles of a given value is equal to 10, replace it with a pebble with a value 10 times larger. Subtraction with such an abacus is similarly simple and obvious.

An abacus is a place-value device, so the Egyptians were certainly aware of the advantages of a place-value number system. Why, during some three millennia, did they never make the transition from an additive to a place-value system? Probably no one ever had the inspiration to invent a zero symbol. Without a place-marker symbol the number 12 3 could be interpreted as 123 with a poorly aligned 3; inserting the zero, as in 1,203, defines the value unambiguously.

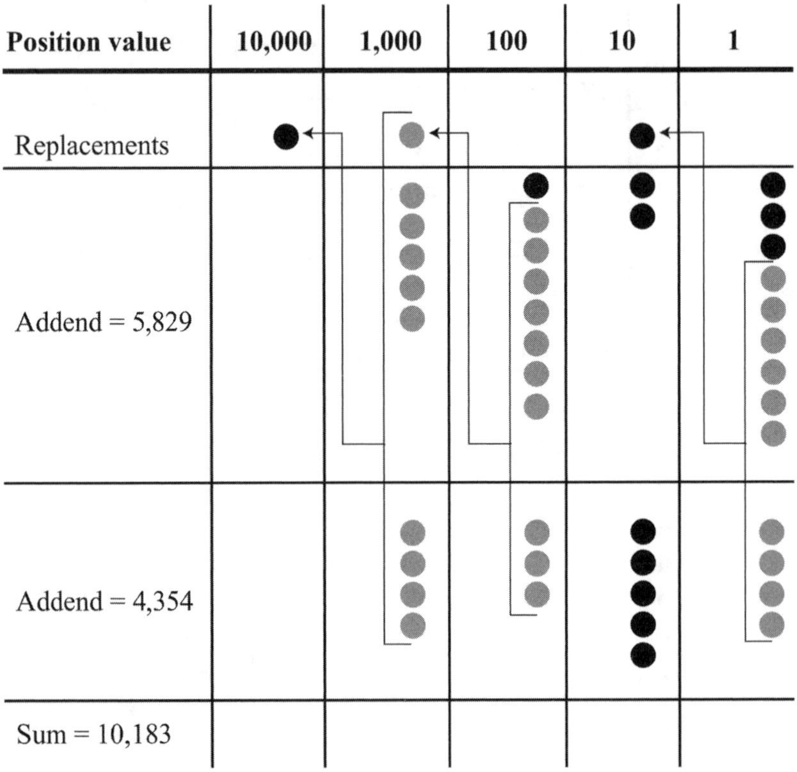

Position value	10,000	1,000	100	10	1
Replacements					
Addend = 5,829					
Addend = 4,354					
Sum = 10,183					

Figure 2.4 Abacus for base-10 addition

Multiplication using multiplication tables would have required even more memorization than using addition tables, but here we have documentation for how the Egyptians solved the problem without any additional memorization burden. Multiplication is just a fast method of successive additions of the same number: $nm = m + m \ldots + m$ (n terms). Calculation by simply adding values of m requires $n - 1$ successive addition operations. Egyptian multiplication very cleverly shortened considerably the number of addition operations by successive doubling. To illustrate the method, consider the multiplication $41 \times m$, where m is any number:

step	operation	\ m
1	$m + m$	$2m$
2	$2m + 2m$	$4m$
3	$4m + 4m$	\ $8m$
4	$8m + 8m$	$16m$
5	$16m + 16m$	\ $32m$ doubling stops because
		$32+32 > 41$
6, 7	$32m + 8m + m$	$41m$

In this example, rather than simply doing $n - 1 = 40$ additions, Egyptian multiplication requires only 7 additions. In Egyptian records, only the last column appears with the entries used in the final summation demarked with the backslash symbol, \ . Successive doubling is equivalent to defining the multiplicand, n (41 in this example), by a base-2 number ($101001 = 2^5 + 2^3 + 2^0$), but it is doubtful that the Egyptians had any awareness of this.

FUN QUESTION 2.1: Using Egyptian multiplication, calculate 83×97. Compare the number of operations required for this calculation with the number of operations required to do a modern pencil/paper calculation with a memorized multiplication table.

FUN QUESTION 2.2: Using Egyptian multiplication, calculate 8051/97. Hint: We, as did the ancient Egyptians, do division by multiplying. Multiplication asks what is x when $N \times M = x$. Division asks what is x when $N \times x = M$, where M and N are known numbers.

When the quotient of a division is not an integer, fractions occur. Consider the division $\dfrac{67}{13}$. In modern notation, we may express the answer either with a common fraction as $5\dfrac{2}{13}$ or with a decimal fraction as $5.\underline{153846}$. The notation of the underlined decimal fraction means that the underlined part is repeated indefinitely. Egyptian fractions have the wonderful property of never being annoyingly nonterminating. In the Egyptian way, $\dfrac{67}{13} = 5 + \dfrac{1}{8} + \dfrac{1}{52} + \dfrac{1}{104}$.

As in this example, all Egyptian fractions are expressed as a finite series of *unit fractions*. As we have just seen, Egyptian arithmetic involves a sequence of doublings. Each doubling operation converts each unit fraction, $1/n$, into a fraction, $2/n$. To facilitate arithmetic with unit fractions the Egyptians composed tables of conversions of $2/n$ into a series of unit fractions. Figure 2.5 is such a table from *RMP*.

n	a	b	c	d		n	a	b	c	d
3						53	30	318	795	
5	3	15				55	30	330		
7	4	28				57	38	114		
9	6	18				59	36	236	531	
11	6	66				61	40	244	488	610
13	8	52	104			63	42	126		
15	10	30				65	39	195		
17	12	51	68			67	40	335	536	
19	12	76	114			69	46	138		
21	14	42				71	40	568	710	
23	12	276				73	60	219	292	365
25	15	75				75	50	150		
27	18	54				77	44	308		
29	24	58	174	232		79	60	237	316	790
31	20	124	155			81	54	162		
33	22	66				83	60	332	415	498
35	30	42				85	51	255		
37	24	111	296			87	58	174		
39	26	78				89	60	356	534	890
41	24	246	328			91	70	130		
43	42	86	129	301		93	62	186		
45	30	90				95	60	380	570	
47	30	141	470			97	56	679	776	
49	28	196				99	66	198		
51	34	102				101	101	202	303	606

Figure 2.5 *Rhind Mathematical Papyrus*—unit fractions

$2/n = 1/a + 1/b + 1/c + 1/d$

The algebraic notation in terms of *n*, *a*, *b*, *c*, and *d* is mine; the Egyptians and the Babylonians had no such notation. Neither did they have operational symbols (+, −, ×, /).

The method of expressing common fractions as a series of unit fractions probably derived from the natural and intuitive use of successive halving. As we have already seen in chapter 1, the most natural fractions are 1/2, 1/4, 1/8, and so on. It was obvious that many frequently occurring common fractions could be expressed as a series of such base-2 unit fractions, $\frac{3}{4} = \frac{1}{2} + \frac{1}{4}$ and $\frac{3}{8} = \frac{1}{4} + \frac{1}{8}$. However, it is not possible to express all common fractions as a finite sum of base-2 unit fractions, and probably the first extension beyond just use of base-2 unit fractions was to allow use of 2/3. Now frequently occurring divisions by a number containing a factor of 3 could also be expressed exactly by a few terms, such as, for example, $\frac{11}{12} = \frac{2}{3} + \frac{1}{4}$. Eventually, to express division by any denominator as a finite series of unit fractions, the Egyptians invented an algorithm that in modern math jargon is called a *greedy algorithm*. We unknowingly and unconsciously use this algorithm when we do pencil/paper long division. Consider a modern pencil/paper division of 67/13 to illustrate use of the greedy algorithm:

- What is the largest (greediest) multiple, n, of 13 such that $13n \leq 67$? $n = 5$
- Remainder $= 67 - 5 \times 13 = 2$
- What is the greediest multiple of 13 such that $13n \leq 2$? $n = 0.1$
- Remainder $= 2 - 0.1 \times 13 = 0.7$
- What is the greediest multiple of 13 such that $13n \leq 0.7$? $n = 0.05, \ldots$
- Thus $\frac{67}{13} = 5.15\ldots$

Now consider Egyptian use of the greedy algorithm to express division of 67/13:

- What is the largest (greediest) multiple, n, of 13 such that $13n \leq 67$? $n = 5$
- Remainder $= 67 - 5 \times 13 = 2 \Rightarrow \frac{67}{13} = 5 + \frac{2}{13}$
- What is the greediest unit fraction that can be subtracted from 2/13?

$$\frac{13}{2} = 6 + \text{ remainder } \Rightarrow 6 < \frac{13}{2} < 7 \Rightarrow \frac{1}{7} < \frac{2}{13} < \frac{1}{6} \Rightarrow \text{greediest unit fraction} = \frac{1}{7}$$

- $\frac{2}{13} - \frac{1}{7} = \frac{1}{91} \Rightarrow \frac{2}{13} = \frac{1}{7} + \frac{1}{91}$

- Therefore $\dfrac{67}{13} = 5 + \dfrac{1}{7} + \dfrac{1}{91}$

This is certainly a simple process, but the Egyptians preferred unit fractions with even denominators because doubling was then simpler. Thus, in the example just given, rather than subtracting the greediest unit fraction, they subtracted something less greedy but even, $\dfrac{2}{13} - \dfrac{1}{8} = \dfrac{3}{104}$, and this process could be continued to obtain a finite sum of unit fractions. Now however, some insightful scribe observed that if he wrote $\dfrac{3}{104} = \dfrac{2}{104} + \dfrac{1}{104}$, then $\dfrac{2}{13} = \dfrac{1}{8} + \dfrac{1}{52} + \dfrac{1}{104}$, which is the entry in *RMP*.

Now let us consider 2/35, also an entry in *RMP*, that was converted into a series of unit fractions just by number smarts. Some scribe assumed that he could express 2/35 by just the sum of two unit fractions, which was reasonable because *RMP* shows that this is true for most small denominators. Letting $\dfrac{2}{35} = \dfrac{1}{a} + \dfrac{1}{b}$ or $2 = \dfrac{35}{a} + \dfrac{35}{b} = \dfrac{5 \times 7}{a} + \dfrac{5 \times 7}{b}$, now he cleverly observed that if $a = 5 \times 6$ and $b = 7 \times 6$, then $2 = \dfrac{7}{6} + \dfrac{5}{6}$ and hence $a = 30$ and $b = 42$.

Clearly, the Egyptians had begun to appreciate some properties of numbers: even and odd, and that some numbers were factorable and some not (prime numbers).

Working with unit fractions was slow but easy and continued to be used for thousands of years. In fact, the clue that conversion of common fractions to unit fractions was by the greedy algorithm was in a textbook by Fibonacci (ca. 1200) that documented the method that was still being used at that time. He was not aware of *RMP* or of anything about Egyptian mathematics; those were all nineteenth-century discoveries. In 1880, Fibonacci's book led the eminent British mathematician James Sylvester to guess that this was also the method of the ancient Egyptians.

FUN QUESTION 2.3: Write the common fraction 1/11 as a sum of unit fractions with even denominators.

Chapter 3

BABYLONIAN NUMBERS
AND ARITHMETIC[1]

B efore about 2000 BCE, Sumer was an important city-state in Mesopotamia. By 2000 BCE, Amorites from the Arabian Desert, by immigration and/or conquest, became dominant. They inherited the Sumerian culture and adopted the Sumerian invention of cuneiform writing to express their own Semitic language. They founded Babylon, which in the reign of Hammurabi (1728–1686 BCE) expanded from a city-state to an empire that dominated all of Mesopotamia.

Before the invention of writing, clay counters (the Sumerian version of pebbles) were used that already exhibited the system of alternating 1-for-10 and 1-for-6 replacements as illustrated in figure 3.1.

After the invention of writing, they made symbols that mimicked the shapes of the counters by pressing round sticks into soft clay. The descriptive content of the tablets on which the numbers appeared allowed unambiguous interpretation of the values of the symbols and hence of the values of the mimicked counters. Figure 3.2 shows the evolution in Sumer from counters to exactly mimicking written symbols to the more efficient written technique of cuneiform. The cuneiform symbols were formed by a succession of indentations made by pressing a wedge-shaped wooden stick into soft clay. As can be seen in the figure, in Sumerian cuneiform the shape of the number symbols continues to mimic the shapes of the clay counters, although somewhat abstractly now.

Like hieroglyphic and hieratic numbers, Sumerian cuneiform numbers were additive, although they were written in an ordered sequence from left to right. Around 1900 BCE, some clever Babylonian scribe realized that Sumerian cuneiform was unnecessarily redundant with the value of a symbol defined by both its unique pattern and its position and he invented a cuneiform place-value number system. Figure 3.3 illustrates how the same number appears in additive Sumerian and in place-value Babylonian.

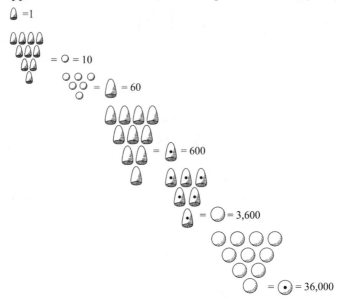

Figure 3.1 Clay counter use in Sumer (ca. 3500 BCE)

	Counters	Written Symbols	
	3500 BCE	3200 BCE	2650 BCE
1	◖	◡	⊻
10	○	○	◁
60	◖	◡	Y
600	◖•	◡•	ᛞ
3,600	○	○	⬠
36,000	⊙	⊙	⬠
216,000	?	?	⬠

Figure 3.2 Evolution of number symbols in Sumer

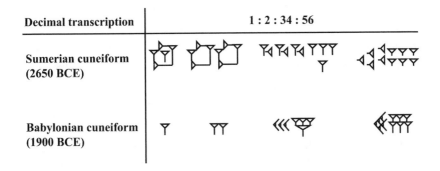

Decimal transcription	1 : 2 : 34 : 56
Sumerian cuneiform (2650 BCE)	
Babylonian cuneiform (1900 BCE)	

Figure 3.3 Transition from Sumerian additive to Babylonian place value

Babylonian numbers from 1 to 59 were now defined by just two symbols: a vertical wedge for a units symbol and a wedge rotated by 90° for a tens symbol. Figure 3.4 presents the cuneiform number symbols for numbers 1 to 59. Vertical wedges additively define numbers 1 to 9. Rotated wedges additively define numbers 10, 20, 30, 40, and 50. There is thus a 1-for-10 replacement within each position and a 1-for-6 replacement from position to position, so that the alternating sequence of 1-for-10 and 1-for-6 replacements is almost equivalent to 1-for-60 position-to-position replacements and hence to a base 60, place-value system.

Figure 3.4 Additive Babylonian cuneiform symbols for numbers 1 to 59

Except for the fact that every position in this base-60 system can have a value up to 59 while every position in the decimal system can only have a value up to 9, cuneiform numbers are theoretically just as easy to read as decimal numbers. But in practice there are difficulties. Using symbols "∨" and "<" as easy-to-type representations of the cuneiform units and tens symbols, the cuneiform number (∨∨ ∨∨∨ <<<<<∨∨) can be decimally transcribed as $2{:}3{:}52_{60}$, which indicates that it is a three-consecutive-positions number but does not unambiguously define what the positions are. We can write the three positions as n, $n + 1$, and $n + 2$, but n is only known from the context of the document, and so its decimal value is only defined as $2 \times 60^{(n+2)} + 3 \times 60^{(n+1)} + 52 \times 60^n$. Assuming n to be zero, then $2{:}3{:}52_{60} = 2 \times 60^2 + 3 \times 60 + 52 = 7{,}432$. However, if the context implied a fraction, then $n = -3$ might be correct and the transcription would be $0.2{:}3{:}52_{60}$ and it would have a decimal value of $2/60 + 3/60^2 + 52/60^3 = 0.034407$.

The Babylonians did not have a sexagesimal point to indicate separation of the integral part of a number from the fractional part; neither did they have a zero symbol. Modern interpreters add these to make the decimally transcribed values unambiguous. One can interpret a number written as (∨∨ <<<<<∨∨) either as $2{:}52_{60}$ or as $2{:}0{:}52_{60}$, because it is not absolutely clear whether the space between adjacent positions is just somewhat larger than usual, or whether an empty position was intended. With Egyptian hieroglyphic or hieratic and Sumerian cuneiform *additive* numbers, a zero symbol had no essential role and this problem did not arise. Egyptian writing also had distinctive symbols for fractions that obviated the need for a decimal point. Thus, prior to the inventions of a zero symbol and a separation point, additive Egyptian hieroglyphic or hieratic and Sumerian cuneiform were easier to read and probably led to fewer mistakes than positional Babylonian cuneiform.

It is difficult to conceive of using our base-10 system without either a zero symbol or a decimal point and thus it appears quite amazing that the Babylonians managed for some 1,500 years with their ambiguously written number system. Eventually, around 500 BCE, the Babylonians did invent a symbol to mark an empty position and thereby invented the zero. It was not until around the first century that "the best of all possible worlds" evolved in India with the combination of the base-10 and positional concepts and eventually after about another millennium reached Europe and became our presently used decimal system.

There is no documentation for how Babylonians did addition and subtraction. Presumably, they used an abacus as in Egypt but designed for the
Babylonian system of alternating replacements of 1-for-10 and 1-for-6 as
illustrated in figure 3.5. For addition/subtraction with an abacus, the Babylonian base-60 number system is marginally more difficult than abacus addi-
tion/subtraction with Egypt's base-10 number system.

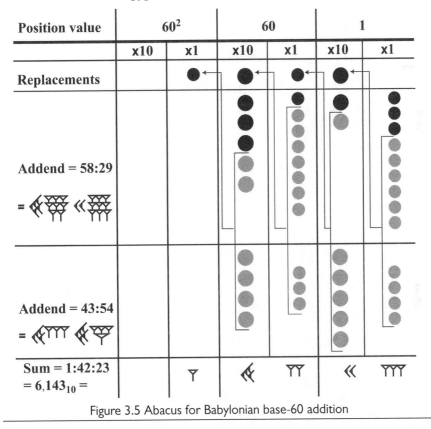

Figure 3.5 Abacus for Babylonian base-60 addition

As noted previously (see chapter 1), if the Babylonian scribes had multiplied using a memorized multiplication table, as we once did before
pencil/paper arithmetic was replaced by electronic calculators, they would
have had to memorize an impossible 1,711 entries. However, they managed to
get by with multiplication tables with fewer entries with the format illustrated
in figure 3.6. Some one hundred such $M \times N$, clay-tablet multiplication tables,
so-called *table texts*, have survived and been recovered. The crude writing and
quality of most of these tablets indicate that they were student-scribe exercises.

Decimal		Decimally transcribed sexagesimal		Sexagesimal cuneiform	
N	10N	N	10N	N	10N
1	10	1	10 (= M)	v	<
2	20	2	20	vv	<<
3	30	3	30	vvv	<<<
4	40	4	40	vvvv	<<<<
5	50	5	50	vvvv v	<<<< <
6	60	6	1:0	vvvv vv	v
7	70	7	1:10	vvvv vvv	v <
8	80	8	1:20	vvvv vvvv	v <<
9	90	9	1:30	vvvv vvvv v	v <<<
10	100	10	1:40	<	v <<<<
11	110	11	1:50	< v	v <<<< <
12	120	12	2:0	< vv	vv
13	130	13	2:10	<< vvv	vv <
14	140	14	2:20	< vvvv	vv <<
15	150	15	2:30	< vvvv v	vv <<<
16	160	16	2:40	< vvvv vv	vv <<<<
17	170	17	2:50	< vvvv vvv	vv <<<< <
18	180	18	3:0	< vvvv vvvv	vvv
19	190	19	3:10	< vvvv vvvv v	vvv <
20	200	20	3:20	<<	vvv <<
30	300	30	5:0	<<<	vvvv v
40	400	40	6:40	<<<<	vvvv <<<< vv
50	500	50	8:20	<<<< <	vvvv << vvvv

Figure 3.6 Cuneiform 10N multiplication table

With this table it was possible to find $M \times N = 10N$ for any $N = 1$ to 59 with no more than two lookups from the table and one addition. For example, 10×37 would have been obtained as $(10 \times 30 + 10 \times 7)$. There are 23 entries in this table, so a set 58 such tables means that the total number of multiplication entries in a complete set of tables is $23 \times 58 = 1,334$, still not a memorizable number of entries. But the table is highly redundant; there is really no need for entries $N = 11$ through 19, and presumably they were included just to illustrate the method of adding entries. The total number of essential entries is now $14 \times 58 = 812$, still not a memorizable number. However, 58 such tables are not necessary. If we allow lookups in two different tables, then only 14 tables are required ($M = 1$ to 9, 10, 20, 30, 40, 50) and there are now only $14 \times 14 = 196$ entries that it would have been necessary to memorize. In fact, examples of all of these 14 table-text multiplication tables have survived. But even this reduced set of tables has redundant entries. If we take, for example, the multiplication of $24 \times 37 = (20 + 4) \times (30 + 7) = 20 \times 30 + 4 \times 7 + 20 \times 7 + 30 \times 4$, we see that all that is required is a units $\times$ units table, a tens $\times$ tens table, and a units $\times$ tens table. Required tables have 36 units $\times$ units entries (see figure 1.1), 15 tens $\times$ tens entries, and 40 units $\times$ tens entries for a total of 91 entries that had to be memorized. The remarkable reduction in required memorization is possible because Babylonian cuneiform is not a true base-60 number system, although in many contexts it can be treated as such, but rather it is a system of alternating 1-or-10 and 1-for-6 replacements. It is possible to reduce the memorization burden further, but there is always a trade-off in more calculation. Ultimately, the multiplication can be reduced to no memorization burden at all by simply adding Ms $(N - 1)$ times.

The surviving student-scribe multiplication tables were clearly exercises in memorizing and using memorized multiplication tables. We have also just seen that it was possible to reduce the memorization burden to a realistic number of entries, but we cannot know exactly how many entries a given scribe remembered. Probably as nowadays, different people remembered different numbers of entries and used different mnemonic devices to fill in missing entries.

The fact that multiplication with memorized multiplication tables was possible and was surely used does not mean that it was the only method used. Egyptian multiplication (see chapter 2) would have been a good method, but no surviving record unambiguously documents its use. Since scribes did parts of calculations that required written steps on soft clay tablets that they then scraped clean to make room for further calculations, there would not have been much written evidence even available to survive. What have survived are some table texts that list successive doublings of numbers. For example, Neugebauer[2] deciphered the table text denoted as CBS 29.13.21 as a sequence of doublings of the regular number 125 (125×2^n, $n = 0$ to 30) and successive halving of its reciprocal ($0.008/2^n$). However, we cannot know if this was for use in Egyptian multiplication, or if it was simply a student exercise in multiplication, or if it was to illustrate a geometric series. As we shall see, there was mathematical interchange between Egypt and Babylon and thus it is probable some Babylonian scribes learned and used Egyptian multiplication.

Mathematicians John O'Connor and Edmund Robertson[3] have stated (ca 2000) that OB scribes invented a multiplication algorithm that required an even smaller memorization burden. The algorithm is equivalent to solving either of the following equations for the product ab.

$$ab = \left[(a+b)^2 - (a-b)^2\right]/4 \qquad (3.1a)$$

$$ab = \left(\frac{a+b}{2}\right)^2 - \left(\frac{a-b}{2}\right)^2 \qquad (3.1b)$$

Multiplication was visualized as calculating the area of a rectangle. These equations calculate the area of a rectangle of length a and width b. The algorithm and geometric visualization that derives equations (3.1) will be given in chapters 4 and 5.

Any number, no matter how large, is multiplied position by position. Since the largest value that any position can have is 59, that is the largest value that $(a+b)/2$ can be. Thus to multiply a by b, using an algorithm that evaluates equation (3.1b), requires having or memorizing a table of squares with just 59 entries. A cuneiform tablet with only 59 entries and with entries on both sides would have a size of about 5 cm by 5 cm (see figure 3.4), about as compact as an electronic pocket calculator.

FUN QUESTION 3.1: When multiplying using equation (3.1b), it is possible to reduce the required number of squares from 59 to only 30. How?

FUN QUESTION 3.2: There is a subtle problem with using equation (3.1b).
- If $(a + b)$ is odd, then $(a - b)$ must also be odd. Prove this.
- If $(a + b)$ and $(a - b)$ are odd, then $(a + b)/2$ and $(a - b)/2$ are not integers and hence cannot be found in tables of squares of integers up to 59^2. Invent a method that enables use of equation (3.1b) with such tables of squares of integers even when $(a + b)$ is odd.

We know that OB scribes used multiplication tables because such tables have survived. We know that OB scribes used tables of squares because such tables have survived, but this does not prove that they used them for multiplication. There are other uses for such tables. As figure 3.7 shows, a table of squares of numbers is also useful as a table of square roots of numbers. Neither does the fact that using tables of squares for multiplication reduces the memorization burden, to a size comparable to that of the present memory burden of our base-10 multiplication table, prove they used tables of squares to multiply.

$m = \sqrt{2}$	2	3	4	5	6	7	8	9	10	11	12	13	14	15
$m^2 = n$	4	9	16	25	36	49	64	81	199	121	144	169	196	225

Figure 3.7 Table of squares and square roots

Since there is no surviving direct evidence for such use, what prompted O'Connor and Robinson to state categorically that OB scribes used tables of squares to multiply? My guess is that their clue was the many surviving examples of use of their algorithm in so-called *problem texts*. But that also is no proof that their algorithm was used in multiplication. There are two points of view.

1. OB scribes first used the algorithm to solve the practical arithmetic problem of multiplication with their base-60 number system, and later

the problem texts were exercises in geometric visualization solutions of quadratic equations. Note that equations (3.1) are quadratic equations.

2. OB scribes used the visualization and algorithm as recreational mathematics, because none of the problem texts in which they used it related to realistic problems. OB scribes may or may not have ever used the visualization and algorithm for practical multiplication.

I prefer the first option; the second option is Hoyrup's.[4] I base my choice on the fact that although there is much commonality between ancient Egyptian and Babylonian mathematics, quadratic geometric algebra problems are missing in the surviving Egyptian mathematical record. To me, this implies that unique OB interest derives from their multiplication problem with a base-60 number system.

One and probably the only advantage of base-60 is that it has more divisors than base-10, thereby allowing simple conversion of some divisions into easier multiplications. However, at best, OB division cannot be simpler than OB multiplication, which as we have just seen is much more awkward than multiplication with base-10.

With any base, division can sometimes produce nuisance, nonterminating fractions. Terminating fractions are comfortable to work with because they are always convertible into integers and are thus essentially equivalent to integers. For example, suppose a measurement yields 3.5 m, a terminating *noninteger*. The measurement in centimeters converts the answer to 350 cm, an *integer*. Without a sexagesimal point and a zero symbol there is no way of distinguishing between an integer and a terminating fraction in OB cuneiform, except from the context of the text.

Egyptian scribes avoided nonterminating fractions with their mixed-base unit fractions (see chapter 2). OB scribes simply avoided divisions that produced nonterminating fractions. OB scribes also constructed tables of *reciprocals* so that multiplication by reciprocals could replace division. The usual criterion for an acceptable division was that the result was either an integer, or equivalently, a terminating fraction. Incidentally, the modern solution to nonterminating fractions is the electronic calculator. It effortlessly calculates nonterminating fractions to far greater precision than is normally required and we can ignore having to decide how many decimal places to keep—at least until the end of the calculation.

In order to understand OB reciprocal tables we must first define some simple, basic properties of numbers. The number 60 is a *composite* number, which means it can be divided into a pair of divisors (in math jargon, it can be *factored*), as, for example, $60 = 6 \times 10$. As shown in the accompanying diagram, a number is factored into a pair of divisors (also called *factors*). Then each of these factors is factored into a pair of divisors and so on until numbers that cannot be factored further are obtained. Numbers that cannot be factored further are *prime* numbers.

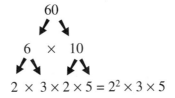

We can generalize from this factorization diagram: every number is either a composite number that can be written as a product of prime numbers or is a prime number. From the diagram, it is clear that the *prime divisors* of 60 are 2, 3, and 5. In math jargon, a number that can be written as $2^i 3^j 5^k$, where i, j, k are integers, is a *regular* base-60 number. Similarly, $2^i 5^k$ is a *regular* base-10 number, and 2^i is a *regular* base-2 number.

If a common fraction, p/q (p and q are integers), has a denominator that is a base-60 regular number, $q = 2^i 3^j 5^k$, and we multiply the numerator and denominator by $30^i 20^j 12^k$ (since $2 \times 30 = 3 \times 20 = 5 \times 12 = 60$), we obtain

$$p/q = p(30^i 20^j 12^k)/60^{(i+j+k)} \tag{3.2}$$

The new denominator determines where to place the sexagesimal point. The new numerator is now the product of integers, a finite number, and hence p/q is a terminating sexagesimal fraction when the denominator, q, is a regular number. We can now tersely restate in math jargon the OB way to avoid non-terminating fractions: divide by *regular* numbers only.

Figure 3.8 presents all of the reciprocals of regular base-60 numbers found in surviving table texts. The columns headed i, j, k, and $2^i 3^j 5^k$ **base-10** are just to make the numbers easier to understand and were certainly not included in OB tables. The columns headed with **base-60** are in decimally transcribed cuneiform.

i	j	k	$2^i3^j5^k$ base-10	$2^i3^j5^k$ base-60	reciprocal base-60	i	j	k	$2^i3^j5^k$ base-10	$2^i3^j5^k$ base-60	reciprocal base-60
0	0	0	1	1	1	5	0	0	32	32	0.1:52:30
1	0	0	2	2	0.30	2	2	0	36	36	0.1:40
0	1	0	3	3	0.20	3	0	1	40	40	0.1:30
2	0	0	4	4	0.15	0	2	1	45	45	0.1:20
0	0	1	5	5	0.12	4	1	0	48	48	0.1:15
1	1	0	6	6	0.10	1	0	2	50	50	0.1:12
3	0	0	8	8	0.7:30	1	3	0	54	54	0.1:6:40
0	2	0	9	9	0.6:40	6	0	0	64	1:4	0.0:56:15
1	0	1	10	10	0.6	3	2	0	72	1:12	0.0:50
2	1	0	12	12	0.5	0	1	2	75	1:15	0.0:48
0	1	1	15	15	0.4	4	0	1	80	1:20	0.0:45
4	0	0	16	16	0.3:45	0	4	0	81	1:21	0.0:44:26:40
1	2	0	18	18	0.3:20	1	2	1	90	1:30	0.0:40
2	0	1	20	20	0.3	5	1	0	96	1:36	0.0:37:30
3	1	0	24	24	0.2:30	2	0	2	100	1:40	0.0:36
0	0	2	25	25	0.2:24	2	3	0	108	1:48	0.0:33:20
0	3	0	27	27	0.2:13:20	3	1	1	120	2:0	0.0:30
1	1	1	30	30	0.2	0	0	3	125	2:5	0.0:28:48

Figure 3.8 Regular base-60 numbers and their reciprocals

Dividing only by regular numbers does not necessarily make division noncumbersome, just less cumbersome. Take, for example, division by 27, a base-60 regular number; whatever number it divides, the result is either an integer or a terminating fraction. Nonetheless, figure 3.8 shows that replacing division by multiplication by a reciprocal still requires multiplication by a three-position sexagesimal number.

How did the scribes identify the regular numbers? Possibly, they simply divided 1 by every number from 2 to 125 and retained for the table only those divisions that produced no more than three fractional positions. Such tedious division may have been the way calculation of reciprocal tables began, but then OB scribes became aware of a much easier method of calculating regular numbers and their reciprocals, which method probably accounts for most of the entries in their reciprocal tables. The clue to the use of this method is a *problem text*, a clay tablet denoted as VAT 6505.

VAT 6505 shows that OB scribes recognized that every large regular number, R, is the product of two smaller regular numbers, $R = R_1R_2$, and that the reciprocal of R can then be calculated by simple multiplication as $1/R = (1/R_1)(1/R_2)$, when R_1 and R_2 and their reciprocals are found in existing reciprocal tables. Therefore, a short table of small regular numbers and their reciprocals, perhaps calculated by tedious division, could be readily extended without limit.

FUN QUESTION 3.3: Prove that if R_1 and R_2 are regular sexagesimal numbers, then $R = R_1R_2$ is also a regular sexagesimal number.

FUN QUESTION 3.4: Is the number 405 a base-10 regular number? Is it a base-60 regular number?

If for base-60 we carry to the limit of the smallest possible regular numbers and recognize that large regular numbers are always products of smaller regular numbers, we obtain that $R = 2^i3^j5^k$, which is just the way I calculated the regular numbers in figure 3.8. If OB scribes also carried their insight to this limit, they would essentially have reached the concept of prime numbers. Whether or not they got that far we do not know, but they certainly at least came very close. Incidentally, as previously noted (see chapter 2), in their calculations of unit fractions, Egyptians of the same era also may have understood the prime number concept. Greek mathematicians of more than 1,500 years later are generally credited with being the first to understand and appreciate the importance of prime numbers.

FUN QUESTION 3.5: How would a lazy OB scribe have divided by 49?

Babylonian understanding of properties of number systems was inherited by succeeding cultures and hence has contributed to the mathematics we use today. But there is one consequence of their mathematics that we celebrate every week. Since seven is a frequently occurring irregular number and division by it was "impossible," seven became an unlucky number and work done on the seventh day was believed to be doomed to failure. The Babylonian solution was to make every seventh day a no-work day. After King Nebuchadnezzar of Babylon conquered them in 604 BCE, the Jews adopted the Babylonian seven-day week, but some imaginative Jewish author gave such adoption a theological spin. He made the seventh day a day of rest, which was required by God's Ten Commandments. The Jewish Sabbath is Saturday. To express their uniqueness, Christians made Sunday their Sabbath. Thus, thanks to Babylonian aversion to irregular numbers and a little theological politics, we now enjoy a two-day weekend.

Chapter 4

OLD BABYLONIAN "QUADRATIC ALGEBRA" PROBLEM TEXTS

I f all surviving OB clay tablets had been carefully excavated from well-defined stratifications, that would certainly have helped establish *provenance* (archaeology jargon for where, when, and by whom). But, as Neugebauer[1] has sadly noted, that is seldom the case. Most artifacts are from the antiquities dealers who have obtained them from natives who were frequently guided unwittingly in their diggings by knowledge of locations of abandoned scientific excavations.

Even if we knew the provenance, that would still not establish when a given calculation was first made because contents of importance were copied and recopied for generations. In this chapter, I discuss four problem texts: YBC 6967, AO 8862, Db_2 146, and VAT 8512. I shall not attempt to date them to greater precision than to say that they are all from the OB era. The sequence is in order of increasing difficulty and sophistication and probably more or less in order of sequence of invention. Thus, I am using only the mathematics itself to establish the dating sequence rather than the more conventional, but less reliable in this case, dating based on linguistic and archaeological evidence. I shall follow the evolution of a type of OB problem text that is equivalent to the solution of quadratic algebraic equations. In parallel I shall also tell the equally interesting story of the evolution of interpretation of such problem texts over the last century.

Figure 4.1 is a drawing of a twenty-first-century BCE clay tablet, MIO 1107, written in Sumerian cuneiform.[2] It records a surveyor-scribe's measurements to determine the area of a field. These data were probably taken for assessing taxes. A quotation from *The Histories*, book II, chapter 109, by the

Greek historian Herodotus (484–430 BCE), although relating to a much later date but to the same inevitable phenomenon, shows that such surveys were also the practice in contemporary Egypt. "Sesostris divided the country among all the Egyptians, giving each man the same amount of land in the form of a square plot. This was a source of income for him, because he ordered them to pay an annual tax. If any of a person's plot was lost to the river, he would present himself at the king's court and tell him what had happened; then the king sent inspectors to measure how much land he had lost, so that in the future the man had to pay proportionately less of the fixed tax." Incidentally, this quotation has always made me wonder whether it implies that every farmer had access to the pharaoh of the Egyptian empire, or if it refers to landowners who were people of wealth. My guess is that reservation of such access to the wealthy is not just in present-day politics.

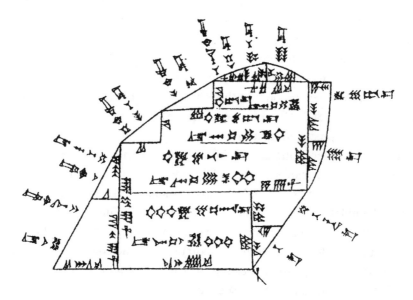

Figure 4.1 Drawing of Sumerian land-survey clay tablet MIO 1107

What is remarkable about this tablet is that it is not remarkable at all. Surveyor data to determine the area of an irregularly shaped field made anytime in the last five thousand years have been similarly recorded and calculated as a sum of areas of rectangles and right triangles. There is no need to assume that this visualization originated in Sumer and was then copied by other cultures. The visualization of an area as a sum of rectangles and right triangles is natural and intuitive. It was probably invented independently in many cultures as a need for land surveys arose. That the area of a rectangle was simply the product of length and width was probably appreciated even before the invention of writing by every farmer who knew that the amount of work required to plow a field increased with the length of the furrows and the number of furrows. In algebraic notation, this logic would be expressed as: if A(area $\propto a$(length) and $A \propto b$(width), then $A \propto ab$. That a right triangle was just a rectangle cut in half by the diagonal was also a natural and intuitive visualization.

Around the twentieth century BCE, observations of such surveyor's patterns may have directed someone's attention to mathematical aspects of patterns produced by division of regular geometric figures. This was a remarkable event. Rumination about such a chance observation was the birth of geometric algebra! In fact, it was probably also the birth of mathematics: it was a generalization, an abstraction, and not just practical arithmetic applied to just simple counting and mensuration.

YBC 6967

"A number exceeds its reciprocal by 7. What are the numbers?"

One of the simplest, possibly the first, and certainly the most important pattern visualized is that of figure 4.2, the formation of two concentric squares by four rectangles. Perhaps the inspiration was the observation of a floor-tile pattern. This is a classic tile pattern that can still be found in catalogs of tile and brick suppliers. (Search under "tile patterns" on the Internet. It is imaginatively called the "windmill" pattern.) Perhaps the observation of construction of a square column with a cross section defined by four rectangular bricks inspired the visualization. This is a classic way of constructing hollow columns and can also be readily observed nowadays on chimneys constructed from cement blocks.

Whatever the inspiration might have been, problems based on this visualization and its variations became an important part of the mathematics curriculum of OB student scribes, as attested by hundreds of surviving problem texts, either as student solutions or as teacher notes. Figure 4.3 is a drawing of problem text YBC 6967. The rough quality of the tablet identifies it as a student's work. I do not know whether the student solved the problem based on previous instruction or if it was a solution dictated by the teacher to teach the method.

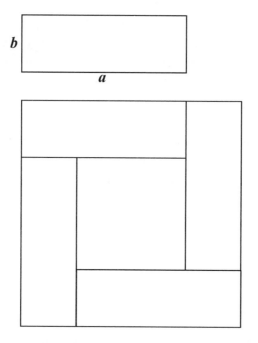

Figure 4.2 Geometric visualization of $4ab = (a + b)^2 - (a - b)^2$

The encircled cuneiform symbols are numbers that you can easily decipher using the definitions given in chapter 3, figure 3.4. In figure 4.4a I have copied Neugebauer's *transcription*, now sometimes more precisely referred to as a *conform transliteration*, in which the cuneiform signs are replaced by their numerical or phonetic values in the same position as the original cuneiform signs. I have omitted some difficult-to-type phonetic symbols, but this transcription will nevertheless give a sense of how the language sounded. I will not be concerned with linguistics questions at this level, except here, to establish the difference between *transcription* and *translation* and to note that almost all of the OB texts I shall consider are based on Neugebauer's transcriptions. There is generally little controversy over a transcription unless the original tablet is badly damaged and/or fragmented and the transcription requires considerable interpolation and extrapolation. Most controversy begins with the translation and increases in the final interpretation because mathematicians and not just linguists can now give their opinions.

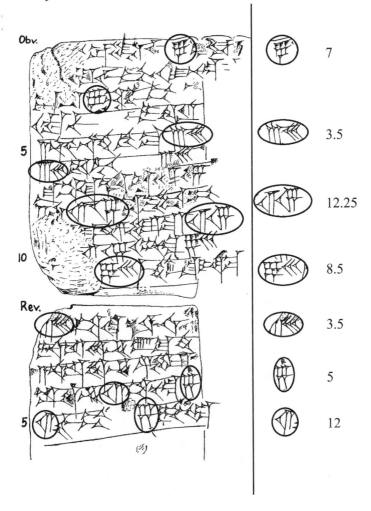

Figure 4.3 Drawing of OB problem text clay tablet YBC 6967

In figure 4.4b, I have copied Neugebauer's translation[3] of YBC 6967, except for my conversion of all numbers to base-10. Italics denote unknown words; square brackets, [ ], denote assumed meaning of illegible text; parentheses, ( ), denote assumed missing words.

Text line	Transcription
1	[igi-b]i e-li igi 7 i-ter
2	[igi] u igi-bi mu-nu-um
3	a[t-t]a 7 sa igi-bi
4	ugu igi i-te-ru
5	a-na si-na he-pe-ma
6	3,30 it-ti 3,30
7	su-ta-ki-il-ma 12, 15
8	a-na 12, 15 sa a-li<-a>-kum
9	[1 eq(?)-1]a-am si-ib-ma 1,12,15
10	[ib-si$_8$ 1],12,15 mi-nu-um 8,30
1 reverse	i-na is-te-en u-su-uh
2 "	i-na is-te-en u-su-uh
3 "	a-na is-te-en si-ib
4 "	is-te-en 12 sa-nu-um 5
5 "	12 igi-bi 5 i-gu-um

Figure 4.4a Neugebauer's transcription of OB problem text YBC 6967

Neugebauer realized that for the text to make sense, the words *igum* and *igibum* were essentially the Babylonian version of a number and its reciprocal. In modern usage, if *a* is a number and *b* is its reciprocal, $ab = 1$, but Neugebauer set $ab = 60$. In chapter 3, we saw that the cuneiform symbol "∨" can be 1 or 60, or any power of 60, so there is some justification in calling the product here a product of reciprocals. Figure 4.5 is my conversion of Neugebauer's translation into its arithmetic algorithm and then into its algebraic generalization.

The visualization of figure 4.2 perfectly accounts for the teacher's composition of the problem. Note that the stated problem is not geometric. What is precocious here is not quadratic algebra as Neugebauer thought, but rather that it was an analysis of a composite geometric figure and an abstraction, a geometric visualization of a nongeometric problem. This same visualization could also have guided the student in logically composing the algorithm rather than just learning a rote solution. In this simple problem, a rote solution was certainly possible, but we shall later see variations that are too complicated for rote solutions to be reasonable. Moreover, what would be the point of a rote solution? This was not a practical problem that a scribe might someday encounter. It was a problem contrived by a teacher-scribe to teach understanding of what he apparently thought was an important mathematics technique. How right he was. The technique was just a hop, skip, and a jump from Euclid, but it took about two thousand years to get there.

Text line	Translation
1	The [*igib*]um exceeded the *igum* by 7.
2	What are [the *igum* and] the *igibum*?
3 – 5	As for you—halve 7, by which the *igibum* exceeded the *igum*, and (the result is) 3.5.
6 – 7	Multiply together 3.5 and 3.5 and (the result is) 12.25
8	To 12.25, which resulted for you,
9	add [60, the produ]ct, and (the result is) 72.25.
10	What is [the square root of 72.25]? (Answer) 8.5.
11	Lay down [8.5 and] 8.5, its equal, and then
1 – 2	subtract 3.5, the *takiltum*, from the one,
reverse	
3 reverse	add (it) to the other.
4 reverse	One is 12, the other 5.
5 reverse	12 is the *igibum*, 5 the *igum*.

Figure 4.4b Neugebauer's translation of OB problem text YBC 6967

The square root in text line 10 is the only potentially difficult arithmetic step. The details of calculating this step are not given and so we can only guess how the student did it. Since the teachers always composed such problems by assuming simple integers for a and b, the result of the student's calculation must also produce simple integers and hence the square root must be of perfect squares. My guess is that the students were taught to do the calculation as $72 \ 1/4 = 289/4 = (17/2)^2$, which could be obtained from a table of perfect squares (see chapter 3, figure 3.7). The last two steps in the algorithm is the solution of a pair of simultaneous linear equations, $(a - b)/2 = 3.5$ and $(a + b)/2 = 8.5$.

Text line	OB algorithm	Algebraic generalization
3 – 5	$7/2 = 3.5$	$(a - b)/2$
6 – 7	$3.5^2 = 12.25$	$[(a - b)/2]^2$
8 – 9	$12.25 + 60 = 72.25$	$[(a + b)/2]^2 = [(a - b)/2]^2 + ab$
10	$\sqrt{72.25} = 8.5$	$(a + b)/2$
3 reverse	$8.5 - 3.5 = 5$	$b = (a + b)/2 - (a - b)/2$
4 reverse	$8.5 + 3.5 = 12$	$a = (a + b)/2 - (a - b)/2$

Figure 4.5 Algorithm and algebraic generalization of OB problem text YBC 6967

This calculation appears to me to be truly algebraic. It is simple enough not to require geometric visualization. I thus partially agree with Neugebauer; this part of the solution was probably algebraic.

From figure 4.2 we can see that there are three ways of composing problems from this visualization:

1. ab and $(a - b)$ given, calculate $(a + b)$, which is YBC 6967.
2. ab and $(a + b)$ given, calculate $(a - b)$, and we shall also see such problem texts.
3. $(a - b)$ and $(a + b)$ given, which is trivially equivalent to a and b given, and no such problem texts exist.

The visualization that the students learned to use to solve the problem is not necessarily the visualization used by the teacher to compose the problem. Since the YBC 6967 algorithm uses $(a - b)/2$ and $(a + b)/2$ rather than just $(a - b)$ and $(a + b)$, this suggests that perhaps the student's visualized drawing may have been figure 4.2 with all dimensions halved. Since the role of the diagram is mnemonic, it really makes little difference whether figure 4.2 or a diagram with dimensions halved was recalled.

Another possible visualization is the right triangle of figure 4.6. This visualization requires knowledge of the Babylonian theorem (defined in the preface) and the Pythagorean theorem. The "proof" that figure 4.2 leads to figure 4.6 will be given in chapter 5, but for the moment let us just accept it.

The right triangle visualization would also serve to remind students of the important Babylonian and Pythagorean theorems. The advantage of this right triangle visualization is that it is documented; most other visualizations are conjectures. No diagrams have ever been recovered that could unambiguously define many of the assumed visualizations, creating an issue for scholars to quibble about for years to come. Lack of surviving diagrams is reasonably, if not completely convincingly, excused by assuming that the diagrams were made on sand tables.

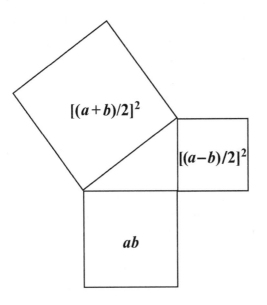

Figure 4.6 Right triangle visualization of OB problem text YBC 6967

FUN QUESTION 4.1: Not only do we not yet know that $(a + b)/2$, $(a - b)/2$, and $\sqrt{ab}$ define a right triangle, we do not even know that they define any triangle. To form a triangle, the sum of the two shortest lengths must be greater than the longest length. Algebraically prove that this is true for these three lengths and hence that they do at least define a triangle.

No one has ever challenged Neugebauer's conclusions regarding his Babylonian arithmetic algorithm or his algebraic generalization of YBC

6967. Without reference to visualization, either figure 4.2 or 4.6, Neugebauer was assuming that OB scribes essentially knew the quadratic equation written on figure 4.2 in rhetorical format, but he never explained how he thought they had arrived at such understanding. My guess is that, for Neugebauer and others who were schooled in symbolic algebra, the ease with which the arithmetic algorithm converts into the algebraic generalization gave rise to an uncritical acceptance of an algebraic derivation.

Such was Neugebauer's stature that his interpretation that OB scribes were doing quadratic algebra to solve this and other problem texts became the gospel spread by his disciples.[4] This was despite the obvious evidence that successor Greek mathematicians did geometric algebra, sometimes referred to as Pythagorean geometric algebra. It is certainly likely that the Greeks benefited from prior Babylonian developments. Biographies of Thales (624–546 BCE) and Pythagoras (580–500 BCE), although possibly more legend than fact, noted that both men had traveled in Egypt and Babylon and this elliptic reference was as far as the Greeks went in acknowledging an intellectual debt to any other culture. Waerden also addressed this point: "Why did the Greeks not simply adopt Babylonian algebra as it was, why did they put it in geometric form?"[5] We now know that his question should have been, "Why did the Greeks turn Babylonian metric geometric algebra into nonmetric geometric algebra?"

How was Neugebauer able to ignore the clue that geometric algebra was known practice in the Greek successor civilization? My guess is that the daunting task of transcribing and translating cuneiform mathematical tablets required so much concentration and effort that he simply did not have sufficient time and energy left to apply to a question of secondary importance. Neugebauer observed that thousands of clay tablets, which had been obtained at huge cost by museums around the world and had been preserved in Mesopotamian soil for thousands of years, were becoming unreadable due to moisture-induced reactions in museum climates.[6] I think that he was single-mindedly dedicated to the translation and the education of a cadre of translators before this irreplaceable material was lost. We can be thankful that he had his priorities right.

At least by 1961, Waerden[7] began to realize that such sophisticated OB algebra without symbolic notation was unlikely, but Aaboe's book that was first published in 1964 remained loyal to Neugebauer's strictly algebraic interpretation. It was not until 1985 that Jens Hoyrup actively challenged Neugebauer's gospel in preprints distributed to colleagues. Hoyrup's geometric algebra visualizations of YBC 6967 and many other problem texts are summarized in his book.[8]

Text line	Translation
1	[The *igib*]*um* over the *ibum*, 7 it goes beyond
2	[*igum*] and *igibum* what?
3	Yo[u], 7 which the *igibum*
4	over the *igum* goes beyond
5	to two break: 3.5
6	3.5 together with 3.5
7	make hold: 12.25
8	to 12.25 which comes up for you
9	[60 the surf]ace append: 72.25
10	[The equalside of 60] what? 8.5
11	[8.5 and] 8.5, its counterpart, lay down.
1 reverse	3.5, the made-hold
2 reverse	from one tear out,
3 reverse	to one append.
4 reverse	The first is 12, the second is 5.
5 reverse	12 is the *igibum*, 5 is the *igum*.

Figure 4.7 Hoyrup's translation of OB problem text YBC 6967

In figure 4.7, I have copied Hoyrup's translation of YBC 6967 word for word, except for my conversion of all numbers to base-10. Whereas Neugebauer translates all operations as arithmetic, Hoyrup interprets some of these operations as descriptions of actions that cut up the *igibum* $\times$ *igum* (*ab*) rectangle and reassemble the pieces as illustrated in figure 4.8. Thus, in text line 5: what Neugebauer translates as "halve," Hoyrup translates as "to two break" and has meaning not only to "halve" but also to break off the piece and paste it back after a rotation of 90° in a different place, which converts the original rectangle into a gnomon ("L" shape) of the same area. The other striking difference is in text line **2 reverse**: What Neugbauer translates as "subtract," Hoyrup translates as "tear out," so that Hoyrup visualizes another "cut-and-paste" operation (not illustrated in figure 4.8) to account for the solution of the pair of simultaneous linear equations.

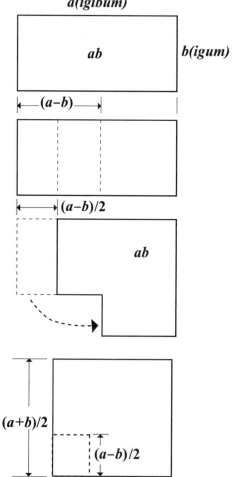

Figure 4.8 Hoyrup's cut-and-paste visualization of OB problem text YBC 6967

Hoyrup's geometric visualization of problem text YBC 6967 does indeed produce a drawing with the required three areas ab, $[(a-b)/2]^2$, and $[(a+b)/2]^2$ (it can be looked at as simply a half-sized figure 4.2 with the central square slid down to the lower-left corner), so it is certainly a possible visualization. However, Hoyrup's translation of text lines **5** and **2 reverse** sound to me like how a Latin-speaking person with imperfect English would translate "subtract" as "to pull out from under." Perhaps when the word "subtract" was coined, the abacus was the calculating device of choice, and "pull out from under" did indeed describe the physical process of subtraction. It is also possible that the actions of Hoyrup's translation also refer to abacus actions rather than geometric actions. Neugebauer[9] also thought that Hoyrup was quibbling over the meaning of synonyms.

But my primary criticism of Hoyrup's visualization is that although it could be a student's solution to a problem composed by a teacher, it is unlikely to have been the visualization that inspired someone to compose the problem. As previously suggested, the inspiration was most likely a chance, insightful observation of a preexisting pattern. That is how such inventions happen, not by virtual cutting and pasting of shapes. Additionally, students would most probably be taught to visualize the same preexisting pattern used by the teacher to compose the problem. Hoyrup's procedure is certainly more complicated than visualizing a preexisting pattern, so there is no advantage in its application for this particular problem.

Interest in such geometric algebra problems began in Old Babylon, continued in ancient Greece, and nowadays high school students enjoy (?) them.

FUN QUESTION 4.2: OB problem text TMS 1 asks: What is the radius of a circle circumscribed about an isosceles triangle whose altitude is 4 units and whose base is 6 units?

FUN QUESTION 4.3: The Greek mathematician Heron (ca. 100 CE) asked: What is the largest square that can be inscribed in an isosceles triangle whose altitude is 4 units and whose base is 6 units?

FUN QUESTION 4.4 The sum of the ages of brothers Pat and Mike is 19. The product of their ages is 60. How old are Pat and Mike? Solve this using the method you learned in high school, not by the Babylonian method. Unless you are an exception, probably the only thing you remember about solving quadratic equations is that there was an equation, "something plus or minus the square root of something." High school mathematics education today, with its emphasis on creating high scores in standardized tests, all to often neglects the derivations where the math is learned and emphasizes memorizing the equations that provide quick solutions in the standardized tests but that are then rapidly forgotten, leaving little appreciation and much distaste for mathematics.

Of course I prefer my visualizations based on the preexisting figures 4.2 or 4.6 to Hoyrup's cut-and-paste visualization of figure 4.8. Thus, I think that at least some of Hoyrup's "quadratic algebra" cut-and-paste visualizations are probably twentieth-century inventions rather than twentieth-century BCE inventions. However, it required someone with linguistics stature to state that the translation required a geometric rather than an algebraic interpretation to get linguistics-oriented scholars to question Neugebauer's quadratic algebra interpretation. Ironically, Hoyrup's cut-and-paste interpretation of YBC 6967, which I think is probably wrong, made geometric algebra interpretations of OB problem texts today's gospel.

As a young graduate student at the Oriental Institute, Oxford University, Eleanor Robson was one of the first after Hoyrup to adopt a geometric visualization, which she used to interpret an OB square root algorithm. I shall discuss this interpretation later in chapter 6. In answer to my question to her of how she felt about challenging Neugebauer's awesome expertise, she wrote, "The short answer to your questions is: ignorance! Neugebauer trained a school of US-based academics. The Assyriologists in late 80s–early 90s Oxford, by contrast, were self-avowedly innumerate—they worked on literary Sumerian, the political history of Assyria, etc. When I came along as a graduate student, their response was, Yes, you can work on maths. We'll help you read the languages, but promise you won't ask us about numbers! So that's how it was. No one even read Neugebauer—it was clearly demarcated as something weird and scary. Jens [Hoyrup]'s article in AOF [*Altorientalische Forschungen*] (1990) came out very soon after I started graduate work, and I just accepted it at face value, because there was no one around to tell me how radical and dangerous it was."[10]

Problem YBC 6967 is the simplest type of problem text that has been called "squares and rectangles." I will consider a few more examples of this type to explore the range of complications the OB teachers dreamed up. As already just seen, it is very difficult to say exactly which visualization is correct; I will choose whichever visualization appears most logical to me, with no implication that it could have been the only OB choice.

AO 8862

"I have multiplied length and width, yielding area. Then I added to the area, the excess of the length over width, yielding 183. Moreover, I have added length and width, yielding 27. Find length, width, area."

AO 8862 is a problem text, translated into German by Neugebauer in the 1930s and into English by Waerden.[11] While YBC 6967 might be logically considered algebraic because the question is not couched in geometric terms, AO 8862 is clearly a geometric problem and geometric algebra interpretation might be expected. Surprisingly, Neugebauer and Waerden considered the solution strong evidence for an algebraic interpretation and hence AO 8862 plays an interesting and important role in the evolution of interpretation of OB problem texts. Rather than considering the student solution to AO 8862, I will consider how the teacher presumably developed the problem. As with YBC 6967, the teacher arbitrarily chose simple integers for dimensions of a rectangle. For YBC 6967 he chose $a = 12$ and $b = 5$; for AO 8862 he chose $a = 15$ and $b = 12$. With reference to figure 4.2, the teacher composed the problem:

$$ab = 180 \tag{4.1a}$$

$$a + b = 27 \tag{4.1b}$$

This problem differs trivially from YBC 6967; $a + b$ is given and it is necessary to find $a - b$ rather than visa versa. Presumably, the students already knew how to solve this problem, so the teacher complicated it a little by adding $a - b = 15 - 12 = 3$ to equation (4.1a) so that the problem became that of AO 8862:

$$ab + (a - b) = 183 \tag{4.2a}$$

$$a + b = 27 \tag{4.2b}$$

Here we have the basis for the Neugebauer/Waerden argument that the solution must be algebraic and not geometric. If the problem is geometric, in equation (4.2a) the term ab has dimensions of area while the term $(a - b)$ has incompatible dimensions of length. Thus, Neugebauer/Waerden concluded that a and b must be viewed as pure numbers and hence the solution is algebraic. The augment can be answered by simply multiplying $(a - b)$ by a length of unity, so that it now becomes an area with the same numerical value. However, I doubt that the OB teacher would have even worried about such an arcane point. The important point here is that even when the problem was geometric, Neugebauer/Waerden viewed the solution as algebraic.

As did Neugebauer/Waerden, I interpret the AO 8862 algorithm as first adding equations (4.2a) and (4.2b), using the same linear algebra operation previously used in the YBC 6967 (figure 4.5, reverse side), which yields equations:

$$ab + 2a = a(b + 2) = 210 \tag{4.3a}$$

$$a + b = 27 \tag{4.3b}$$

Next, another very clever linear algebraic operation was used: $b' = b + 2$, yielding equations

$$ab' = 210 \tag{4.4a}$$

$$a + b' = 29 \tag{4.4b}$$

The remainder of the solution is then just application of the geometric algebra of figure 4.2 with equations (4.4) in mind, which yields $a = 15$ and $b' = 14$, and then one more simple linear algebra step yields $b = b'- 2 = 12$.

FUN QUESTION 4.5: Solve equations (4.4) using the figure 4.2 visualization.

Hoyrup[12] gives a cut-and-paste interpretation of AO 8862, which I again think improbable for the same reasons previously voiced with respect to YBC 6967.

DB$_2$ 146

"The area of a rectangle is 0.75 and its diagonal is 1.25. What is its length and width?"

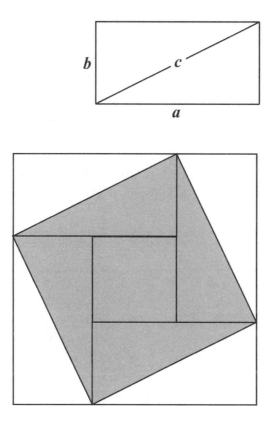

Figure 4.9 Geometric visualization of OB problem text Db$_2$ 146

Problem text Db$_2$ 146 is another interesting variation. It was only found in 1962 and so there is no Neugebauer transcription, translation, and interpretation.[13] Hoyrup presents a translation and interpretation based on visualization of the preexisting diagram of figure 4.9, which is simply figure 4.2 with the rectangles cut on the diagonal and replaced by right triangles.[14] I note that even Hoyrup was unable to justify a cut-and-paste solution for this problem, though he gave a cut-and-paste solution for YBC 6967 from which Db$_2$ 146 was clearly derived. While visualization of a preexisting diagram always provides a solution, cut and paste does not, which is another reason why I doubt its general validity.

The geometry of the shaded part of the diagram is equivalent to the equation

$$c^2 = 2ab + (a - b)^2 \tag{4.5}$$

Since ab and c are givens, $(a - b)^2$ can be calculated. After ab and $(a - b)$ are known, the problem is the same type as YBC 6967 and hence the solution can also be based on the visualization of figure 4.2. Figure 4.10 presents the OB arithmetic algorithm and its algebraic generalization. The OB tablet is rather redundantly and diffusely composed with 25 text lines. To make the solution easier to understand, I have condensed the algorithm to just 10 lines. I refer to each line of calculation simply as "step" to differentiate from Hoyrup's translation, which lists all 25 text lines. Steps 1–5 are based on visualization of figure 4.9; steps 6–10 are based on visualization of figure 4.2.

Step	Babylonian arithmetic algorithm	Algebraic generalization
1	$2 \times 0.75 = 1.5$	$2ab$
2	$(1.25)2 = 1.5625$	c^2
3	$1.5625 - 1.5 = 0.0625$	$(a - b)^2 = c^2 - 2ab$
4	$\sqrt{0.0625} = 0.25$	$(a - b) = \sqrt{(a - b)^2}$
5	$0.25/2 = 0.125$	$(a - b)/2$
6	$0.0625/4 = 0.015625$	$[(a - b)/2]^2$
7	$0.75 + 0.015625 = 0.765625$	$[(a + b)/2]^2 = ab + [(a - b)/2]^2$
8	$\sqrt{0.765625} = 0.875$	$(a + b)/2 = \sqrt{[(a+b)/2]^2}$
9	$0.875 + 0.125 = 1$	$a = (a + b)/2 + (a - b)/2$
10	$0.875 - 0.125 = 0.75$	$b = (a + b)/2 - (a - b)/2$

Figure 4.10 Algorithm and algebraic generalization of OB problem text Db$_2$ 146

When the teacher-scribe composed YBC 6967 and AO 8862, he was free to choose any dimensions for the rectangle and he chose small integers to make the arithmetic simple because the purpose was to teach visualization, not arithmetic. But in Db$_2$ 146 the diagonal is determined by the sides of the rectangle according to the Pythagorean theorem ($a^2 + b^2 = c^2$). Now, in order to keep the arithmetic simple, the triangle's dimensions must be limited to Pythagorean triples, integer solutions to the Pythagorean theorem. The chosen dimensions are defined by the simplest Pythagorean triple (3, 4, 5), but it was common OB practice to divide each dimension by 4 to obtain one dimension of unity, so the assumed dimensions here were (0.75, 1, 1.25).

While it is clear that the teacher was aware of the (3, 4, 5) Pythagorean triple, he may not have been aware of its generalization as the Pythagorean theorem. What is particularly intriguing about Db_2 146 is that it could have been used as a rigorous proof of the Pythagorean theorem, as rigorous as that given some two millennia later by Euclid. There is no way of knowing if they understood this, and I make no claim that they did, but it is interesting to see how this proof could have been done.

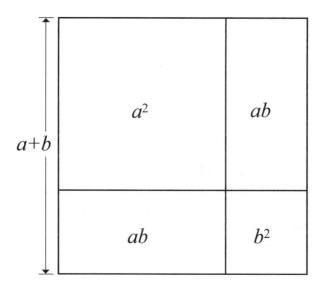

Figure 4.11 Geometric visualization of $(a + b)^2 = a^2 + b^2 + 2ab$

Figure 4.11 is another rectangles-and-squares composite diagram that OB scribes used. It was used in OB square root calculations and its use will be discussed later in chapter 6. This visualization is equivalent to the algebraic equation

$$(a + b)^2 = a^2 + b^2 + 2ab \tag{4.6a}$$

The visualization of figure 4.9, the complete figure and not just the shaded part, is equivalent to the algebraic equation

$$(a + b)^2 = (c^2 + 2ab) \qquad\qquad (4.6b)$$

Equating equations (46a) and (46b) produces the Pythagorean theorem. This is one of the most quoted of the approximately five hundred known proofs and has been attributed to Pythagoras.[15] Incidentally, this proof appears now to be the proof of choice in high school geometry texts because it is so much easier to understand than Euclid's proofs.

However, if teachers and students knew the Pythagorean theorem and the Babylonian theorem when Db_2 146 was composed and presented, then the problem text could have been based only on the visualization of figure 4.6. I do not think that the translation can distinguish between these two alternative solutions.

FUN QUESTION 4.6: Solve Db_2 146 using only the visualization of figure 4.6.

Although I have given much more attention to the solution based on figure 4.9 than to the solution based on figure 4.6, that is just because figure 4.9 presents a new visualization and new concepts rather than because I think it a more probable solution. Neither should the fact that both Hoyrup and I independently proposed the figure 4.9 visualization render that a more probable interpretation. We both had the same clue—awareness of its modern use as a proof of the Pythagorean theorem.

VAT 8512

"Find the dimensions of a right triangle. The base is 30. The triangle is divided by the line x parallel to the base such that $A_1 - A_2 = 420$ and $y_2 - y_1 = 20$." See figure 4.12 for the definition of these algebraic symbols that were surely not in the OB inscription. A more literal translation is just too difficult to understand.

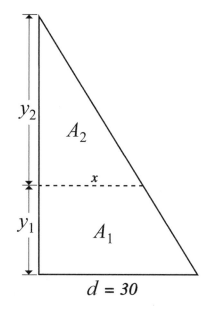

Figure 4.12 Definition of variables for OB problem text VAT 8512

Problem text VAT 8512 was a real brain-teaser, not just for some unfortunate OB student but also for a succession of translators and mathematicians. Neugebauer gave an algebraic solution in the 1930s, Gandz made further progress with a geometric interpretation in 1948, and Huber gave a more comprehensible solution in 1955. Huber's geometric interpretation was apparently what made Waerden recognize the shortcomings of purely algebraic solutions.[16] Perhaps the subtlety of the solution prevented scholars and mathematicians other than Waerden from appreciating the implications of the Gandz/Huber interpretation. My treatment closely follows Waerden. Hoyrup gives a new translation.[17]

The mission of the VAT 8512 problem is to find the dimensions of the right triangle illustrated in figure 4.12. The base is given as 30 units. The triangle is divided by the line x, parallel to the base, into areas A_1 and A_2, and verticals y_1 and y_2, such that $A_1 - A_2 = 420$ and $y_2 - y_1 = 20$. There are three unknowns (x, y_1, and y_2) and hence an algebraic solution requires the solution of the three equations

$$y_1(x + 30)/2 - y_2 x/2 = 420 \tag{4.7a}$$

$$y_2 - y_1 = 20 \tag{4.7b}$$

$$y_2/y_1 = x/(30 - x) \tag{4.7c}$$

Neugebauer had no problem solving these equations algebraically. He apparently did not doubt that the Babylonians were also capable of such complicated algebra.

FUN QUESTION 4.7: Solve equations (4.7) algebraically. It is not simple even with symbolic algebra, so you can appreciate how difficult it would have been with just rhetorical algebra.

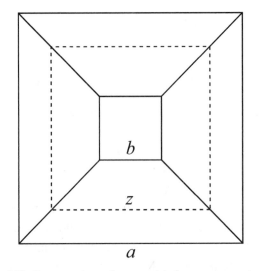

Figure 4.13 Construction of trapezoids from concentric squares

---

I will again start with the teacher's visualization that inspired composing the problem. The apparently unrelated visualization of figure 4.13 shows how the area of a trapezoid can be defined by concentric squares. The difference between the areas of squares of sides a and b define a trapezoid of area

$$A_{ab} = (a^2 - b^2)/4 \tag{4.8a}$$

The difference between the areas of squares of sides a and z define a trapezoid of area

$$A_{az} = (a^2 - z^2)/4 \qquad\qquad (4.8b)$$

The problem, "what is the value of z such that $A_{az} = A_{ab}/2$," is then simply solved as

$$z^2 = (a^2 + b^2)/2 \qquad\qquad (4.8c)$$

Although this derivation is for the specific case of symmetric trapezoids with sides sloping at 45°, surprisingly, the result applies to any trapezoid. We cannot know whether the OB scribe who composed the problem understood this or it was the only solution he knew, and he used it as an approximation. We shall later see in chapters 7 and 11 that approximations were indeed used when exact solutions were not known. However, it is also possible that the scribe applied the following logic that assured that at least he had a reasonable approximation: let $b = 0$ and the trapezoid becomes a triangle and equation (4.8c) still gives the correct answer; let $a = b$ and the trapezoid becomes a rectangle and equation (4.8c) again gives the correct answer.

FUN QUESTION 4.8: Referring to figure 4.12, show that $x = a/\sqrt{2}$ divides the triangle into two equal areas.

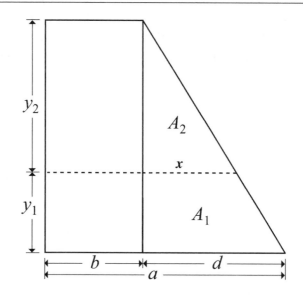

Figure 4.14 Visualization of the solution of OB problem text VAT 8512

Armed with equation (4.8c), the teacher-scribe adds a rectangle to the triangle of figure 4.12 to create the trapezoid illustrated in figure 4.14. He arbitrarily chose the dimensions of the triangle, base = 30, altitude = $y_1 + y_2$ = 40 + 60 = 100, as generally is the case in OB problem texts, to produce simple integer values for all givens and the results of all calculations. Since $y_2 > y_1$, as width b of the rectangle increases, the area above line x increases more than the area below. Since $A_1 > A_2$, for some value of b the area above x will equal the area below x, which the student will calculate as

$$y_2 b + A_2 = y_1 b + A_1 \Rightarrow (y_2 - y_1)b = A_1 - A_2 \Rightarrow b = 420/20 = 21, a = 21 + 30 = 51$$

Applying equation (4.8c):

$$z = \sqrt{(51^2 + 21^2)/2} = 39 \Rightarrow x = z - b = 39 - 21 = 18$$

Applying equations (4.7b) and (4.7c) yields $y_1 = 40$, $y_2 = 60$.

When viewed as a student's solution, the addition of the rectangle to the triangle is too ingenious. When viewed as the teacher's composition using a known property of a trapezoid, equation (4.8c), and dividing the trapezoid into a triangle and a rectangle, the problem is simply clever. Incidentally, even Hoyrup agrees that this problem text is based on the preexisting diagram of figure 4.13, which is itself just a variation of figure 4.2 that Hoyrup prefers to define by a cut-and-paste construction. I find this a hard-to-accept disjoint.

Chapter 5

PYTHAGOREAN TRIPLES

P*ythagorean triples*, or simply *triples*, are sets of three integers (a, b, c) that form right triangles. A likely beginning scenario for the origin of OB awareness of triples is awareness of the (3, 4, 5) triple. The awareness had to begin somewhere and there is no simpler, well-documented candidate than the (3, 4, 5) triple. Chance observation that this set of lengths created a right triangle appears reasonable. In the preceding chapter, we have already seen evidence of its use in OB problem texts Db_2 146 and TMS 1 (see FUN QUESTION 4.2). Problem text Db_2 146 shows that OB scribes also realized that $3^2 + 4^2 = 5^2$ and that $(3/4)^2 + (4/4)^2 = (5/4)^2$.

OB PROBLEM TEXT BM 85196 #9

A much simpler example[1] that can leave no doubt about OB understanding of the properties of the (3, 4, 5) triple is problem text BM 85196 #9. Illustrated in figure 5.1, the problem reads: A ladder of length 30 stands against a wall. The upper end has slipped down a distance 6. How far did the lower end move?

$$\text{lower end moved } \sqrt{30^2 - 24^2} = \sqrt{324} = 18$$

The (18, 24, 30) triple is just a 6 $\times$ (3, 4, 5) triple. The solution shows that OB scribes certainly understood that $3^2 + 4^2 = 5^2$, and also that $(3n)^2 + (4n)^2 = (5n)^2$, where n is any number.

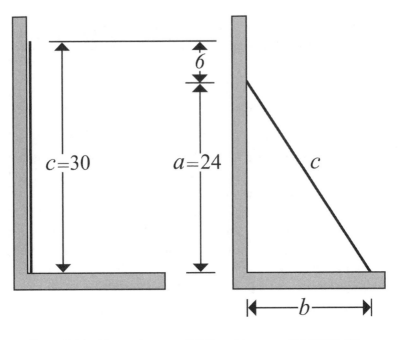

Figure 5.1 Ladder-against-a-wall OB problem text BM 85196 #9

BERLIN PAPYRUS 6610 #1

There is no completely unambiguous evidence of Egyptian awareness of triples or the Pythagorean theorem, but new translations[2] of a papyrus fragment, Berlin Papyrus 6610 #1, dated to about 1300 BCE, conclude that it is a problem text that reads:

Two quantities are given; one is 3/4 of the other.
The sum of the squares of the quantities is 100.
What are the quantities?

Expressed algebraically,

$$b = \frac{3}{4}a \tag{5.1a}$$

$$a^2 + b^2 = 100 \tag{5.1b}$$

The algebraic solution of this simple quadratic-linear pair of equations, as we learned in high school, is to use equation (5.1a) to eliminate b in equation (5.1b), which yields

$$a^2\left[1+\left(\frac{3}{4}\right)^2\right] = 100 \Rightarrow a\sqrt{\frac{25}{16}} = 10 \Rightarrow a = 8 \Rightarrow b = 6$$

Neither the Egyptians nor the Babylonians were capable of even such a simple algebraic solution. But they were both able to solve such problems with the method of "false proposition":

(*assume side is* 1). The false proposition.

(*then the false other side is* $\frac{3}{4}$)

(*then the false diagonal* is) $\sqrt{1+\left(\frac{3}{4}\right)^2} = \sqrt{\frac{25}{16}} = 5/4$

(*true diagonal* = $\sqrt{100}$) = 10

(*correction factor* = *true diagonal/false diagonal*) = 8

(*therefore side is*) $8 \times 1 = 8$ (*and other side* is) $8 \times \frac{3}{4} = 6$

This solution may not be by the simplest possible algebra, but I think it qualifies as algebraic. Elementary algebra is simply generalized arithmetic, thus any algorithm that specifies a defined sequence of arithmetic operations qualifies as algebraic. A solution by simply random guessing would not qualify as algebraic.

We can also now understand why in general, and in particular in problem text Db$_2$ 146 (see chapter 4), the triangle was defined as (0.75, 1, 1.25) rather than as (3, 4, 5); it simplified the algebraic solution.

In the translation of the solution just given, I have enclosed all of the rhetoric in parentheses because it has all been assumed by the translator to make the solution understandable. No word such as *diagonal* that would give a clue that a triangle was visualized appears on the papyrus fragment. Although it is not explicitly stated that the calculation relates to a triangle, when we note the many similarities between Egyptian and OB calculations it is a reasonable guess that this papyrus does relate to the (3, 4, 5) triple and that understanding of the right-triangular properties of (3, 4, 5) was widely understood in the ancient Middle East.

Incidentally, the use of (3, 4, 5) triangles in problems did not go out of fashion some four thousand years ago. We saw an example of its use in first-century Greece (see FUN QUESTION 4.3) and you will find it frequently used in elementary physics books nowadays. Figure 5.2 defines such a typical physics problem. The internal angles in the (3, 4, 5) triangle are 37° and 53°. However, modern usage of the (3, 4, 5) triangle is different than ancient use. The reason for the modern usage is the trigonometric functions are easily remembered and utilized (sin 53° = 4/5, tan 53° = 4/3), while ancient use was usually to provide square roots that could be easily and exactly determined from a table of squares of integers.

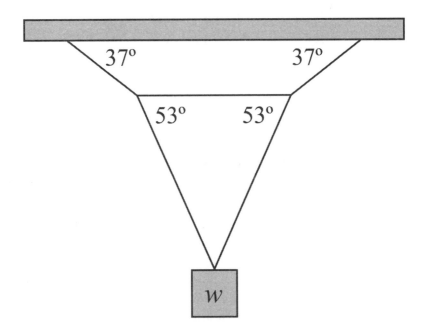

Figure 5.2 A Physics 101 problem

FUN QUESTION 5.1: Do you remember any elementary physics? As the weight in figure 5.2 is increased, eventually one of the ropes will break. Assuming that all ropes are of equal strength, which rope will break first?

Natural curiosity might have led to questioning whether sets of three integers other than just (3, 4, 5) also produced right triangles. Perhaps some scribe simply constructed triangles with other combinations of lengths and discovered that, for example, (5, 12, 13) produced a right triangle and also noted that in this case $5^2 + 12^2 = 13^2$, and then generalized such sparse evidence to conclude: for all right triangles defined by three integers (a, b, c), $a^2 + b^2 = c^2$. Alternatively, perhaps triples other than (3, 4, 5) might have been discovered by chance observations of a table of squares of integers such as figure 3.7, which we know were frequently used. Some clever scribe might have noticed that sometimes the sum of the squares of two numbers was given by the square of a third number: $9 + 16 = 25$, $25 + 144 = 169$ and that the corresponding sets of integers (3, 4, 5) and (5, 12, 13) produced right triangles. He then generalized such sparse evidence to conclude: whenever three integers (a, b, c) satisfy the equation $a^2 + b^2 = c^2$, the integers produce a right triangle.

In sum then, we have no idea which came first, awareness that when three integers (a, b, c) produce a right triangle they satisfy $a^2 + b^2 = c^2$ or awareness that when three integers (a, b, c) satisfy $a^2 + b^2 = c^2$ they produce a right triangle. In fact, we do not even know that these guessing scenarios produced any new triples. They may have, but (3, 4, 5) and (5, 12, 13) that appeared in YBC 6967 are the only triples that are used in any OB problem text.[3] Thus, from problem texts, there is no evidence that Babylonians were aware of any triples other than $(3n, 4n, 5n)$ and $(5n, 12n, 13n)$.

PROTO-PLIMPTON 322

In 1945, Neugebauer[4] published his translation of Plimpton 322, an OB table text of integer solutions to $a^2 + b^2 = c^2$. However, most of the sets of three integers, (a, b, c), in Plimpton 322 were huge; the largest set was (12,709; 13,500; 18,514). The only method available to OB scribes to check whether sets of three integers were triples was to actually construct or draw the triangle and test whether it was indeed a right triangle, but this was impossible for such large numbers. Thus, there must have been some calculation prior to Plimpton 322, which I will call proto-Plimpton 322, that produced enough integer sets that satisfied $a^2 + b^2 = c^2$ but were small enough to construct or draw triangles, and thus "prove" that any set of three integers that satisfies $a^2 + b^2 = c^2$ is a triple. It was not a rigorous Euclidian proof of the Pythagorean theorem, but it was a reasonable proof, and it was sufficient for OB scribes henceforth to apply the Pythagorean theorem to any right triangle.

There is no surviving text that exhibits this "proof," but we can guess how it was done using the visualization of problem text YBC 6967 in figure 4.2, from which it rigorously follows that

$$(a+b)^2 = (a-b)^2 + \left(2\sqrt{ab}\right)^2 \qquad (5.2)$$

where a and b are the sides of a rectangle. If we draw the diagonal, then a and b are the sides of a right triangle. We have already noted (see FUN QUESTION 4.2) that each term in equation (5.2) can be interpreted as the square of a side of a triangle. If we want these three sides to be integers, then a and b must be integers and the product, ab, must be a perfect square. For example, if we choose $a = 4$ and $b = 1$, we obtain the ever-popular (3, 4, 5) triple as was illustrated in figure P.1 in the preface.

In order to be able to use the standard notation of (a, b, c) for the calculated triangles and to avoid confusion with the a and b of equation (5.2) that refer to the input data and not to the output of three integers, I will change the symbols in equation (5.2) to P and Q. Equation (5.2) now becomes

$$(P+Q)^2 = (P-Q)^2 + \left(2\sqrt{PQ}\right)^2 \qquad (5.2a)$$

$$c^2 \quad = \quad a^2 \quad + \quad b^2$$

One simple way of generating sets of three integers with equation (5.2a) is to let P be a perfect square and $Q = 1$. Figure 5.3 is my calculation of triples using this method.[5] I can now refer to them as triples because the values of (a, b, c) are small enough to test and they all construct right triangles. We have now "proved" the Babylonian theorem and the Pythagorean theorem, and presumably, this calculation or some variation of it was also the OB "proof." We have seen that a rigorous general proof, of which even Euclid would have approved, could have been derived from the visualization of Db$_2$ 146 (see chapter 4), but we cannot be sure that anybody in Old Babylon had the mathematical insight to see that. The best bet is that OB justification for use of the Pythagorean theorem came from such proto-Plimpton 322 calculations.

P	Q	$P + Q$	$P - Q$	$2\sqrt{PQ}$	internal angle	internal angle	triple (a, b, c)
4	1	5	3	4	36.87	53.13	3, 4, 5
9	1	10	8	6	36.87	53.13	3, 4, 5
16	1	17	15	8	28.07	61.93	8, 15, 17
25	1	26	24	10	22.61	67.39	5, 12, 13
36	1	37	35	12	18.92	71.08	12, 35, 37
49	1	50	49	14	16.25	73.75	14, 49, 50
64	1	65	63	16	14.24	75.76	16, 63, 65
81	1	82	80	18	12.67	77.33	9, 40, 41
100	1	101	99	20	11.42	78.58	20, 99, 101

Figure 5.3 Proto-Plimpton 322: Pythagorean triples

FUN QUESTION 5.2: Using equation (5.2a), derive two triples for $Q \neq 1$.

Equation (5.2a) is valid for any values of P and Q, and not just for when PQ is a perfect square. When PQ is not a perfect square, solution of equation (5.2a) requires an ability to calculate square roots, which was a problem in the ancient world, even into the Hellenistic times. OB calculations of square roots will be considered in chapter 6.

Although no OB documentation has been found for most of the proto-Plimpton 322 triples presented in figure 5.3, pre-Pythagoras use is clearly documented in the Hindu Sulbasutra that possibly was written as early as 800 BCE but was probably known much earlier. The Sulbasutra describes construction of sacrificial alters with shapes defined by triples, as presented in Figure 5.4.[6] Note that all of the triples employed are found in proto-Plimpton 322. That in itself is not compelling evidence that the calculation reflects OB origins. However, when we consider the Sulbasutra method of graphically calculating square roots presented in figure 5.5, we see that it is obvious use of the Babylonian theorem. The Sulbasutra instructs how to square a rectangle, which is just another way of saying to calculate a square root. Using the notation of equation (5.2) to express the rhetorical algebra of the Sulbasutra, for a rectangle of sides a and b, construct the line $(a - b)/2$. At one end of $(a - b)/2$ erect a perpendicular and using the other end as a center draw an arc of a circle of radius $(a + b)/2$. Where the arc cuts the perpendicular defines a length equal to $\sqrt{ab}$. I consider the combination of figures 5.4 and 5.5 strong evidence that sometime in the interval 2000 to 800 BCE the Hindus learned the "proofs" of the Babylonian and Pythagorean theorems from the Babylonians.

Capital letters at the corners in figure 5.4 are anachronistically added to unambiguously and unobtrusively define the construction. It was not Hindu practice but a Greek invention, probably in the fifth century BCE, that was instrumental in enabling mathematical progress. I use it anachronistically to define pre-Greek geometry, just as I anachronistically use modern algebraic notation. This *Greek alphabetic annotation* will be considered in more detail in subsequent chapters.

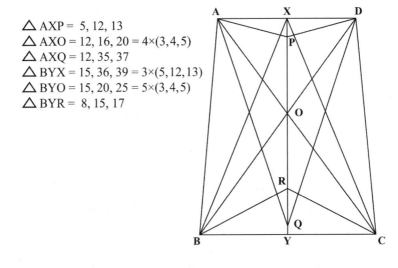

$\triangle$ AXP = 5, 12, 13
$\triangle$ AXO = 12, 16, 20 = 4×(3,4,5)
$\triangle$ AXQ = 12, 35, 37
$\triangle$ BYX = 15, 36, 39 = 3×(5,12,13)
$\triangle$ BYO = 15, 20, 25 = 5×(3,4,5)
$\triangle$ BYR = 8, 15, 17

Figure 5.4 Hindu sacrificial altar defined by Pythagorean triples

It is interesting to note that the translation of the Sanskrit word "Sulba" is "rope" and thus the Sulbasutras are "rope rules." The alter design is described as stretching a rope with knots at one-foot intervals to define the various triples. Possibly this was also the OB way of testing that the results of the proto-Plimpton 322 calculations really did produce triples.

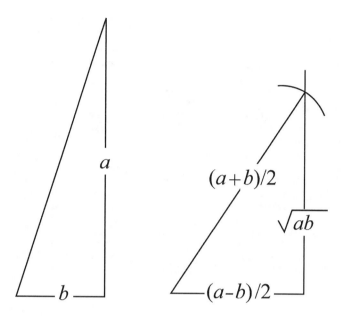

Figure 5.5 Hindu calculation of square root using the Babylonian theorem

PLIMPTON 322

Now let us return to Plimpton 322, which can now be easily understood as a small evolutionary step from the proto-Plimpton 322 calculation. Figure 5.6 is a drawing of Plimpton 322. It is a clay tablet some 12.7 by 8.8 cm. The high quality of the writing shows that it was the work of a professional and not a student. It is easy to discern that it is a table of columns of numbers. Most obvious is column IV, which is just a listing of row numbers, 1–15, and is of no further interest. The left side of the tablet is broken off somewhere in column I. This has created a small enigma about what exactly were the values in column I and a big enigma that has generated reams of speculations about what the missing part might have contained.

Col. I   Col. II   Col. III   Col. IV

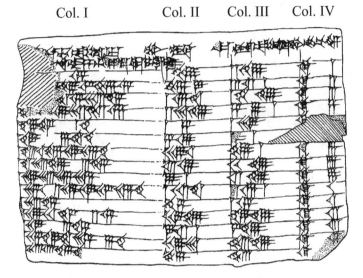

Figure 5.6 Drawing of OB table text Plimpton 322

Neugebauer[7] uncannily saw that if he squared the value in column III and subtracted from it the square of the value in column II, he obtained a number that was itself a square of an integer. Neugebauer had discovered that Plimpton 322 was a table of *Pythagorean triples*, integer solutions to the Pythagorean theorem, $a^2 + b^2 = c^2$. The Plimpton 322 way of making PQ into a perfect square was to make both P and Q perfect squares individually, which simply means that $P = p^2$ and $Q = q^2$, where p and q are any integers. Thus, the equation for generating triples now evolved to equation (5.2b):

$$\left(p^2 + q^2\right)^2 = \left(p^2 - q^2\right)^2 + \left(2pq\right)^2 \tag{5.2b}$$

$$c^2 \quad = a^2 \quad\quad + b^2$$

and its use is now frequently referred to as the *pq* method.

The OB scribe who composed Plimpton 322 added one other refinement that takes into account the uniquely OB problem of division with their base-60 number system. He restricted values of p and q to frequently used regular numbers. Chapter 3 noted that OB scribes would usually only divide by regular numbers because division by irregular numbers produced nuisance, nonterminating fractions. The values of *p* and *q* used in Plimpton 322 are limited to regular numbers previously given in figure 3.8.

Figure 5.7 presents my transcription into modern base-10 notation of Plimpton 322. It is not the transcription of the entries as they appear on the clay tablet but rather after Neugebauer and others have corrected arithmetic and copying errors and filled in illegible text with logical values. I have added the shaded areas in the table to make the interpretation easier. I do not presume that any of these shaded-area entries represent anything in the missing part of the tablet to the left of column I, although some might. The table gives Pythagorean triples that cover the range from about 30° to 60° by limiting the range of *p/q* to $1.8 \leq p/q \leq 2.4$. Presumably, two other similar tables, which have not been found, spanned the ranges 0° to 30° and 60° to 90°.

p	*q*	*p/q*	Slope (degrees)	Col. I $(c/b)^2$	$b = 2pq$	Col.II $a = p^2 - q^2$	Col.III $c = p^2 + q^2$	Col. IV
12	5	2.4	44.8 or 45.2	[1].983	120	119	169	1
64	27	2.37	44.2 or 45.8	[1].949	3,456	3,367	4,825	2
75	32	2.34	43.8 or 46.2	[1].919	4,800	4,601	6,649	3
125	54	2.31	43.3 or 46.7	[1].886	13,500	12,709	18,514	4
9	4	2.25	42.1 or 47.9	[1].815	72	65	97	5
20	9	2.22	41.5 or 48.5	[1].785	360	319	481	6
54	25	2.16	40.3 or 49.7	[1].720	2,700	2,291	3,541	7
32	15	2.13	39.8 or 50.2	[1].692	960	799	1,249	8
25	12	2.08	38.7 or 51.3	[1].643	600	481	769	9
81	40	2.03	37.4 or 52.6	[1].586	6,480	4,961	8,161	10
2	1	2.00	36.9 or 53.1	[1].563	(60) 1	(45) 3/4	(75) 5/4	11
125	64	1.95	35.8 or 54.2	1.519	16,000	11,529	19,721	
48	25	1.92	35.0 or 55.0	[1].489	2,400	1,679	2,929	12
15	8	1.88	33.9 or 56.1	[1].450	240	161	289	13
50	27	1.85	33.3 or 56.7	[1].430	2,700	1,771	3,229	14
9	5	1.8	31.9 or 58.1	[1].387	90	56	106	15

Figure 5.7 Translation of Plimpton 322

The calculations almost perfectly account for Plimpton 322. The row between tablet rows 11 and 12 is the only *pq* pair in the *p/q* range of the other entries in the tables that was presumably missed by whoever wrote or copied the tablet. Only in tablet row 11 is there a major discrepancy. This triple that has been transcribed as (45, 60, 75), which uses the same convention for transcribing cuneiform numbers as the other entries in the tablet, is apparently anomalously 15 × (3, 4, 5).[8] However, as noted in chapter 3, there is always some ambiguity in transcribing cuneiform numbers and so tablet row 11 could also have been transcribed as (3/4, 1, 5/4), which is just a (3, 4, 5) triangle normalized to $b = 1$, which as we have already seen was the preferred way the triple was used in problem texts in both Egypt and Babylon. This usage gives us a clue to the reason why the composer of Plimpton 322 chose regular numbers for p and q. The normalization divides each side of the triangle by $b = 2pq$ and hence it makes this calculation much easier when b is a regular number (see chapter 3).

The mystery is not why row 11 is normalized but why the other rows are not. However, this mystery is not important because OB tablets are not like present-day publications that should be completely intelligible to the designated reader and should not contain such unexplained inconsistencies. The tablets were either reminders to the writer himself or school exercises accompanied by drawings and oral explanations. Perhaps the normalization of just one line was to remind the author of a better way to present the data, and there are no doubt other just-as-plausible explanations.

FUN QUESTION 5.3: In the transcription of Plimpton 322 of figure 5.7, except for the (45, 60, 75) triple, when you normalize by dividing by $b = 2pq$, a regular number, why do you get a nonterminating fraction? Why do you get a terminating fraction when you normalize the (45, 60, 75) triple?

FUN QUESTION 5.4: In the transcription of Plimpton 322 of figure 5.7, if you normalize any of the huge triangles, you obtain small triangles. Why does that not convert Plimpton 322 into a practical trigonometric table?

The intended purpose of Plimpton 322 is still a mystery.[9] My guess is that the intent was to compose a trigonometric table to define useful triangles, probably for use in construction. However, most of the triples turned out to be impractically large. Except for the (3, 4, 5) triple, no other triple in Plimpton 322 has ever turned up in any OB document.

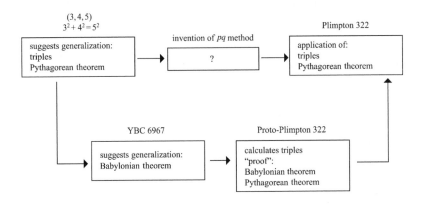

Figure 5.8 Alternative paths to Plimpton 322

Figure 5.8 diagrams the way I think that the evolution of Plimpton 322 should be regarded. The "classic" way requires an independent derivation of the *pq* method. Neugebauer slyly avoided this question altogether. He simply showed that modern understanding of the *pq* method accounts for Plimpton 322. Waerden fell into the trap that Neugebauer had avoided and attributed Neugebauer's derivation to the Babylonians.[10] He essentially adopted a derivation by Diophantus of Alexandria (ca. 250). Waerden starts with assuming the Pythagorean theorem, which he rearranges and factors:

$$b^2 = c^2 - a^2 = (c - a)(c + a) \tag{5.3}$$

Dividing equation (5.3) by b^2 yields a product of a pair of reciprocals:

$$\left(\frac{c}{b} + \frac{a}{b}\right)\left(\frac{c}{b} - \frac{a}{b}\right) = 1 \tag{5.4}$$

Since a, b, and c must all be integers, both of the factors in equation (5.4) contain only common fractions and the factors themselves must then also be given as common fractions:

$$\left(\frac{c}{b} + \frac{a}{b}\right) = \frac{p}{q} \tag{5.5a}$$

$$\left(\frac{c}{b} - \frac{a}{b}\right) = \frac{q}{p} \tag{5.5b}$$

where p and q are integers.

Now, first adding and then subtracting equations (5.5a) and (5.5b), we obtain

$$\frac{c}{b} = \frac{\left(\dfrac{p}{q} + \dfrac{q}{p}\right)}{2} \tag{5.6a}$$

$$\frac{a}{b} = \frac{\left(\dfrac{p}{q} - \dfrac{q}{p}\right)}{2} \tag{5.6b}$$

$$\left(\frac{c}{b}\right)^2 - \left(\frac{a}{b}\right)^2 = \Rightarrow a^2 + b^2 = c^2 \tag{5.6c}$$

So using equations (5.6a) and (5.6b) in equation (5.6c) generates equation (5.2b), which enables calculation of the triples in Plimpton 322. This is a very sophisticated, algebraic derivation. Diophantus lived after more than seven hundred years of Greek mathematical developments. His special interest was solution of indeterminate equations, equations with many possible integer solutions, and the Pythagorean theorem is such an equation. That this could have been the OB solution is highly unlikely.

FUN QUESTION 5.5: Derive equation (5.3) using the cut-and-paste visualization of figure 4.8. That it can be derived by this visualization does not mean that it was. Robson has proposed such a derivation.[11] Even if equation (5.3) could have been geometrically derived, there still remain too many sophisticated algebraic steps for this to have been the OB derivation of the *pq* method.

Plimpton 322 is certainly one of the most important ancient mathematical texts deciphered in the twentieth century. Although Neugebauer's insightful translation had surprisingly revealed OB understanding of the Pythagorean theorem from more than a millennium before Pythagoras, its importance in twentieth-century BCE mathematics was probably minimal.

Chapter 6

SQUARE ROOT CALCULATIONS

EGYPTIAN CALCULATION

To calculate $\sqrt{2}$, the Egyptians apparently applied the observation that an area could be approximately doubled if its linear dimensions were increased by a factor of $7/5 = 1.4$ (see chapter 1, figure 1.6). They apparently understood that an area of any shape is proportional to some linear dimension squared and that $\sqrt{2} \cong \frac{7}{5} = 1\frac{2}{5} = 1.4$, derived from $2 = \frac{50}{25} \cong \frac{49}{25}$. This is apparently the only ancient Egyptian example of a square root of a nonperfect square and even it could only be calculated by approximating it by a ratio of perfect squares.

OB PROBLEM TEXT YBC 7289

OB scribes had a better approximation for easy use in normal calculations, $\sqrt{2} \cong 1\frac{5}{12} = 1.417$. They also did a surprisingly precise calculation of $\sqrt{2}$. Figure 6.1a is a drawing of OB clay tablet YBC 7289 that defines this precise calculation. Figure 6.1b presents the numbers in decimally transcribed cuneiform. As usual, there is some ambiguity about absolute values of cuneiform numbers. The self-consistent choice I use is: side of square = $0.30_{60} = 0.5_{10}$; diagonal (number under horizontal diagonal) = $0.43{:}25{:}35_{60}$

= 0.707107 ($\cong \sqrt{0.5}$); the number above the diagonal = $1.24{:}51{:}10_{60}$ = 1.414213 is the OB calculation of $\sqrt{2}$, to six-digit precision.

($\sqrt{2}$ = 1.414213562 to the ten-digit limit of precision of my electronic calculator.)

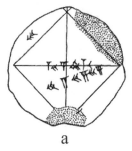

a

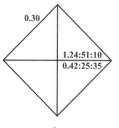

b

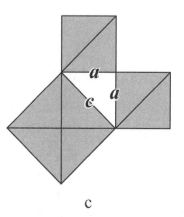

c

Figure 6.1 Drawing of OB problem text clay tablet YBC 7289

The motivation for doing this extremely tedious calculation is a mystery. In the surviving record at least, not only is this result never used again, the algorithm is never again applied to calculate precisely the square root of any other number. Nor was there any practical need for such precision in a mud-brick civilization. Thus, the reason for the calculation was presumably simply mathematical curiosity. Curiosity about what?

As noted in chapter 3 and in the discussion of Plimpton 322 in chapter 5, the Babylonians divided numbers into regular and irregular and preferred to use regular numbers because division by irregular numbers produced nuisance, nonterminating fractions. My guess is that $\sqrt{2}$ mystified them. Was it regular or irregular? The only way they knew to answer this was to calculate and see if the result terminated. It is doubtful that they were mathematically insightful enough to be able to see that their algorithm to calculate $\sqrt{2}$ would never terminate and hence that $\sqrt{2}$ was an irregular number.

One of the most important and sophisticated Greek proofs was that $\sqrt{2}$ is irrational, which means that it cannot be expressed as a common fraction p/q, where p and q are integers. (Because of its importance here and later in this book, this proof is given in detail in appendix C.) It is doubtful that any OB scribe had any inkling of the concept of irrationality, but perhaps some realized that there was something peculiar about $\sqrt{2}$. Division by irregular numbers never terminates, but the sequence of digits is cyclic (for example, in base-10: $1/11 = 0.090909\ldots$); for irrational numbers the sequence of digits is not cyclic. However, no example has yet been found of OB division by an irregular number that had been carried out to a sufficient number of digits to exhibit cyclic behavior, so it is doubtful that OB scribes were aware of the property.

Two diagonals divide the square drawn on YBC 7289 into four triangles. Presumably, the drawing on the tablet is an incomplete copy of some diagram such as figure 6.1c, which shows that the sum of the areas of squares on the two sides of the triangle (a, a, c) is equal to four triangles. This is also the area of the square with its side defined by the diagonal, the triangle (a, a, c). This complete diagram was possibly another "proof" of the Pythagorean theorem, albeit only for the special case where the two sides of a right triangle are equal, $2a^2 = c^2$.

OB SQUARING-THE-RECTANGLE (Heron's Method)

Now let us try to find a visualization and algorithm that can account for the precise OB calculation of $\sqrt{2}$. We cannot be sure how the approximations for $\sqrt{2}$ were calculated, but it is logical to look for a visualization and its algorithm that first yields the usual approximation and then can be continued to yield the precise calculation.

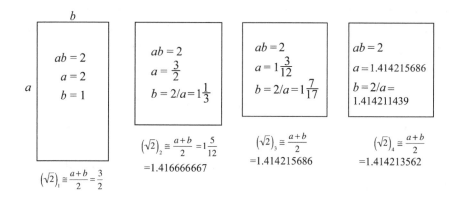

Figure 6.2 Squaring the 2 x 1 rectangle

Figure 6.2 illustrates a possible natural and intuitive visualization, squaring of a rectangle. We start with an $ab = 2 \times 1$ rectangle. Clearly, the average of the two sides yields a value closer to a side of a square, so it is a natural and intuitive choice. Thus, we obtain as a first approximation

$$a_1 = \frac{a+b}{2} = \frac{2+1}{2} = \frac{3}{2} = 1.5$$

For the next approximation, since $a_1 b_1 = 2$, $b_1 = \frac{4}{3}$, so that for the second approximation,

$$a_2 = \frac{a_1 + b_1}{2} = \frac{\frac{3}{2} + \frac{4}{3}}{2} = \frac{17}{12} = 1\frac{5}{12} = 1.417$$

which is just the usual OB approximation noted previously. The calculation continues similarly with the rectangle becoming progressively squarer,

$$a_3 = \frac{a_2 + b_2}{2} = \frac{\dfrac{17}{12} + \dfrac{24}{17}}{2} = \frac{577}{408} = 1.414216 \text{ and}$$

$$a_4 = \frac{a_3 + b_3}{2} = \frac{\dfrac{577}{408} + \dfrac{816}{577}}{2} = \frac{665,857}{470,832} = 1.414213562$$

which is correct to ten-digit precision. Visualizing squaring of a rectangle and using this iterative algorithm yields both the usual OB approximation and the precise OB calculation and thus could have been used.

An OB approximation for $\sqrt{3}$ is also known,[1] $\sqrt{3} = 1.45_{60} = 1\frac{3}{4}$. This can also be derived by the squaring-the-rectangle visualization and algo-

rithm: $3 \times 1 \Rightarrow 2 \times \frac{3}{2} \Rightarrow \dfrac{2 + \dfrac{3}{2}}{2} = \frac{7}{4} = 1.75$ (my electronic calculator gives

$\sqrt{3} = 1.732050808$).

This algorithm, which is sometimes credited to Heron of Alexandria (ca. 100), was proposed by Neugebauer.[2] His derivation was typically purely algebraic and without any geometric visualization. The logic of his derivation is that if a first guess for $\sqrt{N}$ is N_1 that errs by being too large (small), then N/N_1 will err by being too small (large), so a better approximation is the average

$$N_2 = \frac{N_1 + N / N_1}{2}$$

The general term of the iterative algorithm can thus be written as

$$N_{n+1} = \frac{N_n + N / N_n}{2} \tag{6.1}$$

This algorithm is exactly the same as the squaring-the-rectangle algorithm used just previously to calculate $\sqrt{2}$ precisely. It is just written in a different algebraic notation.

Surviving copies of Heron's works show that he actually did use the squaring-the-rectangle visualization, not Neugebauer's algebraic derivation.

Neugebauer did not claim that some OB scribe had used his derivation but only that his algorithm accounted for the data. There may have been a subliminal hint that an OB derivation resembled his derivation, but in his usual sly style, Neugebauer avoided suggesting how an OB scribe might have derived the result so he could never be accused of speculation unsupported by textual evidence.

OB CUT-AND-PASTE SQUARE ROOT (Newton's Method)

Another natural and intuitive visualization also could have been used. Consider the first diagram in figure 6.3. The large square has a known area of 2, but we do not know the length of the side, $\sqrt{2}$. Inside this square is a square with a known side. I have chosen a value of $\frac{3}{4}$, essentially using the same starting guess as in figure 6.2 since $\frac{4}{3} \times \frac{3}{2} = 2$. Figure 6.3 shows that the square with an area of 2 is the sum of the area of the inside square and the area of the gnomon (shaded area) of width ε,

$$2 = \left(\frac{4}{3}\right)^2 + 2\left(\frac{4}{3}\varepsilon\right) + \varepsilon^2$$

With the approximation that neglects the small ε^2 term, this equation can be solved for ε,

$$\varepsilon = \frac{1}{12} \Rightarrow \sqrt{2} \cong \frac{4}{3} + \frac{1}{12} = 1\frac{5}{12}$$

exactly the same result obtained using squaring-the-rectangle visualization and algorithm.

If rather than taking 4/3, that is, less than $\sqrt{2}$, as the first guess, we had taken 3/2, then we would have obtained

$$\varepsilon = \frac{1}{12} \Rightarrow \sqrt{2} = \frac{3}{2} - \frac{1}{12} = 1\frac{5}{12}$$

the same result as previously. Thus, it makes no difference if the first guess is less than or greater than the number whose square root is being calculated. This is just an academic point because it is doubtful that negative numbers were appreciated in the OB era.

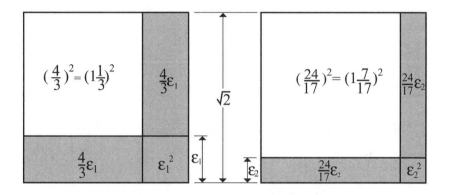

Figure 6.3 OB visualization for calculating $\sqrt{2}$

Figure 6.3 also shows that the next step also gives the same result as the next step in the squaring-the-rectangle algorithm. Thus, although starting from different visualizations, both algorithms give exactly the same results. Thus, as far as calculations of $\sqrt{2}$ and $\sqrt{3}$ are concerned, there is no way to decide which visualization and algorithm OB scribes used.

When used as an iterative algorithm, the general term can be more elegantly written as

$$a^2 = (a_n + \varepsilon_n)^2 \cong a_n^2 + 2a_n\varepsilon_n \implies \varepsilon_n = \frac{a^2 - a_n^2}{2a_n} \tag{6.2a}$$

$$a_{n+1} = a_n + \varepsilon_n \tag{6.2b}$$

Using the right-hand equation (6.2a) in equation (6.2b) yields

$$a_{n+1} = \frac{a_n + a^2 / a_n}{2} \tag{6.2c}$$

which is just equation (6.1) in geometric terms.

This iterative algorithm was invented by Isaac Newton (1642–1726) and hence is sometimes referred to as Newton's method.

Figure 6.4a shows how this square root calculation can also be visualized as a cut-and-paste procedure. To find the square root of the sum of two areas a^2 and B, when $B \ll a^2$ ($\ll$ means much less than), the area B is visualized as spread out as a rectangle along one side of the larger square so that $B = 2a\varepsilon$. Then this rectangle is sliced in half, and half is pasted along the other side of the larger square.

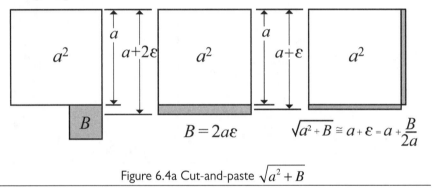

Figure 6.4a Cut-and-paste $\sqrt{a^2 + B}$

Since
$$a^2 + 2a\varepsilon + \varepsilon^2 = (a + \varepsilon)^2 \tag{6.3a}$$

and using $\varepsilon = B/2a$, we obtain
$$a^2 + B + \varepsilon^2 = \left(a + \frac{B}{2a}\right)^2 \tag{6.4a}$$

Neglecting the small ε^2 term and taking the square root of equation (6.4a),
$$\sqrt{a^2 + B} \cong a + \frac{B}{2a} \tag{6.5a}$$

which is obviously just the first step in the iterative algorithm defined by equations (6.2). Taking the existence of the ε^2 term into account when taking the square roots of equation (6.4a),
$$\sqrt{a^2 + B} < a + \frac{B}{2a} \tag{6.6a}$$

The reasoning behind this conclusion is that in equation (6.4a), the left-hand side (LHS) equals the right-hand (RHS) side. When we neglect the ε^2 term in the LHS, LHS < RHS.

Figure 6.4b is the cut-and-paste visualization to find the square root of the difference between two areas a^2 and B, when $B \ll a^2$. The area B is visualized as spread out as a rectangle along one side of the larger square so that $B = 2a\varepsilon$. Since

$$a^2 - 2a\varepsilon + \varepsilon^2 = (a - \varepsilon)^2 \qquad (6.3b)$$

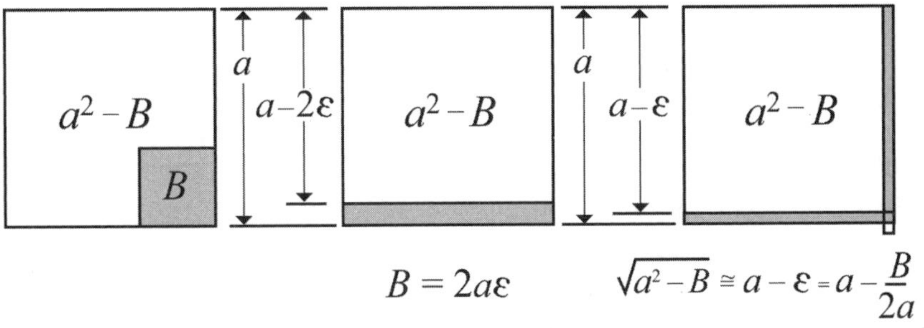

$$B = 2a\varepsilon \qquad \sqrt{a^2 - B} \cong a - \varepsilon = a - \frac{B}{2a}$$

Figure 6.4b Cut-and-paste $\sqrt{a^2 - B}$

and using $\varepsilon = B/2a$, we obtain

$$a_2 - B + \varepsilon^2 = \left(a - \frac{B}{2a} \right)^2 \qquad (6.4b)$$

Neglecting the small ε^2 term and taking the square root of equation (6.4b),

$$\sqrt{a^2 - B} \cong a - \frac{B}{2a} \qquad (6.5b)$$

Taking the ε^2 term into account when taking the square root of equation (6.4b),

$$\sqrt{a^2 - B} < a - \frac{B}{2a} \qquad (6.6b)$$

The inequalities expressed in equations (6.6) play an important role in Greek square root calculations, hence this detailed treatment of what might be considered a trivial point.

The calculation visualized in figure 6.3, which yields equations (6.2), is not obviously cut-and-paste. But since the calculation is identical to the cut-and-paste calculations of figures 6.4 and I cannot think of a good name for the figure 6.3 visualization, I shall simply refer to both as cut-and-paste.

Figure 4.11 in chapter 4 that could have enabled interpreting problem text Db$_2$ 146 as a proof of the Pythagorean theorem is this very same visualization. Thus, if the scribe who composed problem text Db$_2$ 146 also knew this cut-and-paste square root visualization, he possibly realized that he had a general, rigorous proof of the Pythagorean theorem.

OB PROBLEM TEXT VAT 6598

Pythagorean theorem calculations documented in OB problem text VAT 6598 are evidence for OB use of equations (6.5).[3]

FUN QUESTION 6.1: A problem in VAT 6598 reads, "*What is the length of the diagonal of a door of height 40 and width 10?*" A student scribe some four thousand years ago could solve this. Can you? Use the cut-and-paste visualization.

FUN QUESTION 6.2: A door has a height of 5 and a diagonal of 22. What is its width? Use the cut-and-paste visualization.

I have devoted considerable attention to the iterative algorithms for calculating square roots, what I have called squaring-the-rectangle (Heron's method) and cut-and-paste (Newton's method) algorithms. They are interesting mathematics and are a significant part of modern interpretation of OB and Greek mathematics, but the evidence that either was ever an OB method is fragile. The iterative algorithms make the precise calculation of $\sqrt{2}$ much easier, but they are not essential. A sequence of guesses can also eventually home in on the same result.

Since the squaring-the-rectangle algorithm gives a natural and intuitive choice for the next guess using the well-documented OB procedure of taking averages, my guess is that it probably was used. I have previously noted a squaring-the-rectangle solution in ancient India (see chapter 5, figure 5.5) and that Heron's visualization was exactly this OB visualization. In chapter 13 we shall see Euclid's squaring-the-rectangle solution. All are evidence for continuing ancient interest in this problem. Only the one-step approximation embodied in equations (6.5), the first step in the cut-and-paste algorithm, has solid OB documentation (VAT 6598 referred to in FUN QUESTION 6.1). As we shall shortly see, only this one-step approximation appears in Greek mathematics from Pythagoras to Ptolemy, so it appears to me as unlikely that the cut-and-paste iteration algorithm was used in the OB era, and thus it was invented, not reinvented by Isaac Newton. The extension of the one-step, cut-and-paste calculation into an iterative algorithm, which with perfect hindsight appears so obvious, was apparently not so obvious.

PYTHAGORAS CALCULATES SQUARE ROOTS[4]

Attributing any specific discovery to Pythagoras (580–500 BCE) is speculative. What is probably true is that he founded a religious cult, the Pythagorean Brotherhood, whose motto was "All is number." They believed that deeper understanding of mathematics would lead to a better life. Despite an overlay of moralistic and mystic nonsense (for example, the number 10 was particularly important because the *tetraktys*, $1 + 2 + 3 + 4 = 10$, represented the four elements from which the universe is made: fire, water, air, and earth), Pythagoreans did discover many important mathematical concepts, which they credited to their revered founder. Thus, when I credit a discovery to Pythagoras, I am just following a tradition of crediting early Greek mathematics to Pythagoras. The extent to which this represents reality is unknown.

Legend has it that Pythagoras traveled to both Egypt and Babylon, and the square root calculation credited to Pythagoras is an interesting combination of OB-era Egyptian and Babylonian calculations. To what extent Pythagoras consciously used prior Egyptian and/or Babylonian inventions is unknown. He used an approximation to define a lower limit to $\sqrt{2}$, which was shown at the beginning of this chapter to have been known to the Egyptians:

$$\sqrt{2} = \sqrt{\frac{50}{25}} > \sqrt{\frac{50-1}{25}} = \frac{7}{5} = 1.4$$

Using $\sqrt{2} = \sqrt{\frac{49+1}{25}}$, and now applying equation (6.5a) to the numerator, with $a^2 = 49$ and $B = 1$, Pythagoras obtained an upper limit of $\sqrt{2}$ using the inequality of equation (6.6a):

$$\sqrt{2} < \frac{1}{5}\left(a + \frac{B}{2a}\right) = \frac{1}{5}\left(7 + \frac{1}{14}\right) = \frac{7}{5} + \frac{1}{70} = \frac{495}{350} = 1.414$$

Thus, Pythagoras determined that

$$\frac{7}{5} < \sqrt{2} < \frac{7}{5} + \frac{1}{70} \tag{6.7}$$

Greek mathematicians were generally not interested in precise numerical values. For their proofs, definitive limits were more useful than precise approximations. Heath gives an example of such use of the lower limit in a proof by Aristarchus (ca. 260 BCE).[5] That $\frac{7}{5} < \sqrt{2}$ is obvious from its derivation, but how did Pythagoras know that $\sqrt{2} < \frac{7}{5} + \frac{1}{70}$? The obvious answer is that he simply did the numerical calculation and found that $\left(\frac{495}{350}\right)^2 > 2$. However, with the Greek number system (similar to the Roman number system we are all familiar with as inscribed dates on pretentious buildings, but with Greek instead of Latin letters) this was a very tedious calculation, and Greek mathematicians generally scorned doing arithmetic. Since $\left(\frac{495}{350}\right)^2 = 2.0002$, the inequality is therefore not clear without meticulous calculation. To me this implies that Pythagoras did not simply accept the approximate equation (6.5a) from a Babylonian source, but he knew and probably had proved the inequality of equation (6.6a).

The limits given in equation (6.7) are somewhat misleading because the upper limit is a good approximation to $\sqrt{2}$, while the lower limit is a poor approximation. Thus, the intuitive averaging of the limits to obtain the best estimate of $\sqrt{2}$ is not applicable here. The best estimate is the upper limit.

ARCHIMEDES CALCULATES SQUARE ROOTS

Until Archimedes (287–212 BCE), there is no record of more precise Greek calculations of square roots than that attributed to Pythagoras. There was just no need for more precise values until Archimedes required them in his study *Measurement of a Circle*,[6] which we shall consider in chapter 7. Archimedes determined that

$$\frac{1,351}{780} > \sqrt{3} > \frac{265}{153}$$

He left no record of how he arrived at this result, so we have to guess. Since it is of the same format given by Pythagoras, he presumably used the same method, just more precisely.

Heath reviews some rather complicated scholarly analyses,[7] but I think that a simpler algorithm is more probable, although nobody can know for sure what Archimedes was thinking. Consider the calculation of $\sqrt{3}$ using equation (6.6b) and the identity $3 = 48/16$

$$\sqrt{3} = \frac{1}{4}\sqrt{7^2 - 1} < \frac{1}{4}\left(7 - \frac{1}{14}\right) = \frac{7}{4} - \frac{1}{56} = \frac{388}{224} = 1.7321$$

Archimedes did better, and in order to understand his method, we must first generalize this numerical calculation. In the identity $3 = 48/16$ both numerator and denominator are perfect squares, let them be a^2 and n^2, respectively, and they must also satisfy the relationship

$$a^2 \pm m = 3n^2 \tag{6.8}$$

In the spirit of the Archimedes calculation, a, m, and n are all integers.

To find a better approximation for $\sqrt{3}$, as presumably did Archimedes, we just systematically assume values for n and look for solutions of equation (6.8) with large a and small m. Figure 6.5 summarizes the search procedure. The calculations of relevance here are in the column titled "UPPER LIMIT." In each calculation of $\sqrt{3}$, I have rounded the number off to the first digit that deviates from the correct answer. It is readily seen that $n = 15$ and $a = 26$ give Archimedes' result. Using equation (6.6b), the calculation is simply

$$\sqrt{3} = \frac{1}{15}\sqrt{26^2 - 1} < \frac{1}{15}\left(26 - \frac{1}{52}\right) = \frac{1,351}{780} = 1.732051$$

Archimedes may have run out of patience at $n = 15$, but my computer program did not and for $n = 56$ and $a = 97$, $m = -1$, equation (6.6b) yields

$$\sqrt{3} = \frac{1}{56}\sqrt{97^2 - 1} < \frac{1}{56}\left(97 - \frac{1}{194}\right) = \frac{18,817}{10,864} = 1.73205081$$

with nine-digit precision since my electronic calculator with ten-digit precision gives,

$$\sqrt{3} = 1.732050808$$

An advantage of my solution is that it is computer programmable, although that was probably not a concern of Archimedes. However, it is computer programmable because it proceeds by a defined algorithm rather than just by clever guessing, and that would have been useful for Archimedes also. This method may be capable of precise calculation, but it is certainly not an efficient method. It requires some fifty-six steps to accomplish what the Heron/Newton method can accomplish in just four steps. It is somewhat mind-boggling to realize that an OB scribe had done a more precise square root calculation than Archimedes, one of the greatest-ever mathematicians.

n	$3n^2$	a	a^2	m	$\left(a+\dfrac{m}{2a}\right)n$	$\left(a+\dfrac{m}{2a+m}\right)n$	$\sqrt{3}\cong\dfrac{UL+3/UL}{2}$
					UPPER LIMIT	LOWER LIMIT	
1	3	2	4	−1	1.75	1.66	1.7321
4	48	7	49	−1	1.7321	1.731	1.732050808
5	75	9	81	−6	1.733		
6	108	10	100	8	1.733		
7	147	12	144	3	1.7321		
8	192	14	196	−4	1.7321		
9	243	16	256	−13	1.7326		
10	300	17	289	11	1.7323		
11	363	19	361	2	1.732057		
12	432	21	441	−9	1.7321		
13	507	23	529	−22	1.7324		
14	588	24	576	12	1.7321		
15	675	26	676	−1	1.732051	1.7320505	1.732050808
.	.	.	.	.	.	.	
.	.	.	.	.	.	.	
56	9,408	97	9,409	−1	1.7320508	1.7320503	1.732050808
			$\sqrt{3}=$		1.7320050808		

Figure 6.5 Calculations of $\sqrt{3}$ upper and lower limits—Archimedes' algorithm

We now have accounted for the upper limit in Archimedes' calculation of $\sqrt{3}$. However, reasoning behind his lower limit is not obvious and scholars have worried over it for millennia. Archimedes determined the upper and lower $\sqrt{3}$ limits as

$$a + \frac{B}{2a+1} < \sqrt{a^2 + B} < a + \frac{B}{2a} \text{ and} \tag{6.9a}$$

$$a - \frac{B}{2a-1} < \sqrt{a^2 - B} < a - \frac{B}{2a} \tag{6.9b}$$

We thus obtain for Archimedes' calculation of the lower limit, corresponding to his best upper-limit calculation,

$$\sqrt{3} = \frac{1}{15}\sqrt{26^2 - 1} < \frac{1}{15}\left(26 - \frac{1}{51}\right) = \frac{1,325}{765} = \frac{265}{153} = 1.732026$$

The source of Archimedes' lower-limit solution, $2a \rightarrow 2a \pm 1$, is a mystery. Did he rigorously derive it or was he simply copying from some prior source? He never even states explicitly that he used it. It is only assumed by modern scholars because it perfectly accounts for his 265/153 lower-limit calculation. Heath notes that it must have been in wider Greek use because the same approximation appears in eleventh-century Arabic mathematical literature that drew from ancient Greek sources.[8] Archimedes might have been the origin of this wider Greek usage.

That Archimedes' lower limit is less than his upper limit is obvious, but that it is lower than $\sqrt{3}$ is not. And no one has ever offered a convincing rigorous proof that it is. My numerical calculations presented in figure 6.5 show that it is, but can we rigorously prove it? Referring to equation (6.9), we want to prove that

$$a - \frac{1}{2a-1} < \sqrt{a^2 - 1} \tag{6.10a}$$

We have already proved (see equation 6.6b) that

$$a - \frac{1}{2a} > \sqrt{a^2 - 1} \tag{6.10b}$$

To prove equation (6.10a), consider the statement

$$\left(a - \frac{1}{2a}\right)\left(a - \frac{1}{2a-1}\right) \leq (a^2 - 1) = 3n^2 \qquad\qquad (6.10c)$$

This equation states that the product of Archimedes' upper limit and lower limit is less than or equal to $\sqrt{3}$. Since we know that the first term on the LHS is greater than $\sqrt{3}$, the only way that the inequality or equality expressed by equation (6.10c) can occur is if the second term on the LHS is sufficiently less than $n\sqrt{3}$. Simply performing the algebraic operations of equation (6.10c) reduces it to

$$1 - a \leq 0 \qquad\qquad (6.10d)$$

which is always true since a is always greater than one and hence we have rigorously proved that Archimedes' lower limit is always less than $\sqrt{3}$. The behaviors of Archimedes' upper and lower limits are schematically illustrated in figure 6.6. At any precision of calculation, the lower limit is almost the mirror image of the upper limit. However, the lower limit is due to an intentionally introduced error and at any degree of precision, as the data in figure 6.5 show, the upper limit alone is still Archimedes' best approximation for $\sqrt{3}$.

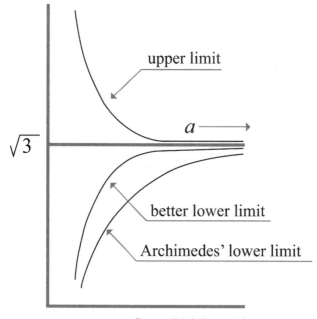

Figure 6.6 (schematic)
Behavior of Archimedes' upper and lower limits in calculations of $\sqrt{3}$

Archimedes' lower-limit calculation was correct, but it was overkill; it was not necessary to make the lower limit so low. He should have calculated the lower limit, *LL*, as

$$UL \times LL = 3 \Rightarrow LL = 3/UL$$

where *UL* is the upper limit that Archimedes had already correctly calculated. This *LL* calculation that is exactly the mirror image of the *UL* calculation is denoted in figure 6.6 as the better lower limit. Then, the average would have been a better approximation than either *LL* or *UL*,

$$\sqrt{3} \cong \frac{UL + 3/UL}{2} \tag{6.10e}$$

The last column in figure 6.5 shows that this calculation is a remarkable improvement. So much so in fact that it obviates the need for most of Archimedes' tedious calculations. We have essentially applied the OB squaring-the-rectangle procedure that had been used more than a millennium before Archimedes.

Archimedes knew that the accuracy of his calculation was dependent on *B*/2*a* since the essence of his precise calculation was to minimize *B* (to 1) and maximize *a*. He just did not have the mathematical tools to do a proper error analysis. But thanks to Isaac Newton, we do. Using Newton's binomial expansion, we obtain

$$\sqrt{3} = \left[(a^2 - 1)^{\frac{1}{2}} \right] / n = \left[a - \frac{1}{2a} - \frac{1}{8a^3} - \dots \right] / n \tag{6.11}$$

For a calculation using *a* = 26 and *n* = 15, the three-term expression of equation (6.11) gives $\sqrt{3}$ precise to ten digits, while Archimedes' calculation or the *a* = 26 and *n* = 15 gives $\sqrt{3}$ precise to seven digits, hence the term $1/8a^3$ is the major part of the error in Archimedes' best upper-limit calculation. It is far less than the error defined by his lower-limit calculation. Archimedes' lower limit is not wrong, as proved above. It is less than $\sqrt{3}$. It is just counterproductive in obtaining a good approximation for $\sqrt{3}$.

When required to produce a better numerical calculation, Archimedes did. It was clever, but it was not particularly original or particularly good. It was just a copycat version of Pythagoras's solution with some improvements. Greek mathematicians just did not like to do arithmetic; they were more interested in discovering eternal and universal truths.

FUN QUESTION 6.3 Prove equation (6.10d).

PTOLEMY CALCULATES SQUARE ROOTS

Ptolemy (83–116 CE) has the dubious honor of being more responsible than any other astronomer for the retention of the concept of an earth-centric universe. This had the effect of legitimizing Catholic dogma, with enormous political and social consequences still felt today. The Greek astronomer Aristarchus (ca. 260 BCE) had in fact proposed the idea of a heliocentric universe a millennium and a half before Copernicus. But because Ptolemy was such a good mathematician, he was able to invent a complicated mathematical model to reconcile observed planetary motion with an earth-centric universe. Ironically, the Catholic Church was so pleased with the unwitting support of Ptolemy that it became a patron of the work of astronomers Copernicus and Galileo, which eventually disproved the earth-centric myth and undermined church influence.

Ptolemy required precise calculations of square roots for his astronomy calculations and essentially invented the method used in recent years when pencil/paper calculations were still being done. Nowadays we use the much more convenient method of pressing the key with the $\sqrt{}$ icon on our ever-handy electronic calculators. I have previously presented this pencil/paper calculation in base-10 in detail,[9] but now I will present it essentially as done by Ptolemy in the peculiar number system used by Greek astronomers.[10] The Greeks had largely adopted the Egyptian number system, base-10 for the integer part of a number, but the fractional part of Egyptian numbers (see chapter 2) was difficult to use. Thus, Ptolemy and other Greek astronomers adopted a mixed-base number system, base-10 for the integer part and the Babylonian base-60 for the fractional part. They obviously realized the advantage of using a number system with a base, and the next logical step would have been to adopt base-10 for both the integer and the fractional parts of a number, but the surprising reality is that this did not happen until a millennium later.

For example, Ptolemy found the square root of 4,500 by applying the *greedy algorithm* (see chapter 2). He sought to express the square root as:

$$integer + \frac{x}{60} + \frac{y}{60^2}$$

He first found the largest perfect square less than 4,500, $67^2 = 4,489$, so that

$$4,500 = 67^2 + 11 = \left(67 + \frac{x}{60} + \frac{y}{60^2} \right)^2 \Rightarrow$$

$$11 = \frac{2 \times 67x}{60} + \frac{x^2}{60^2} + \frac{2 \times 67y}{60^2} + \frac{2xy}{60^3} + \frac{y^2}{60^4} \qquad (6.12a)$$

The largest possible integer value of x is 4 for the first and largest term on the RHS of equation (6.12a) to be less than 11. Inserting this value for x in equation (6.12a), he obtained

$$11 - \frac{536}{60} - \frac{16}{60^2} = 3 - \frac{56}{60} - \frac{16}{60^2} = \frac{2 \times 67y}{60^2} + \frac{8y}{60^3} + \frac{y^2}{60^4} \qquad (6.12b)$$

The largest integer value of y is 55 for the first and largest term on the RHS of equation (6.12b) to be less than the LHS. In fact for $y = 55$, equation (6.12b) is adequately satisfied and thus

$$\sqrt{4,500} = 67 + \frac{4}{60} + \frac{55}{60^2} = 67.0819 \qquad (6.12c)$$

$\sqrt{4,500} = 67.0820$, according to my electronic calculator, rounded off to six figures.

Ptolemy's square root calculation is not only more original, it is better than Archimedes'. Ptolemy was undoubtedly an exceptional mathematician and astronomer, but he epitomizes science at its worst. Because we observe planetary motion from a moving earth, the motion is apparently complex. Ptolemy accounted for the complex motion, assuming a fixed earth by what amounts to just mathematical curve fitting with as many adjustable parameters as required. The fact that he obtained a reasonable fit says nothing about reality. Work like his inspired Occam's razor, William of Ockham's (1285–

1349) criterion for good science, which in Albert Einstein's (1879–1955) version is "… make the irreducible basic elements as simple and as few as possible without having to surrender the adequate representation of a single datum of experience," often more pithily expressed as *"make it simple, not simpler."*

FUN QUESTION 6.3: Show that equation (6.12b) is satisfied with $y = 55$, without converting to base-10.

FUN QUESTION 6.4: Using Ptolemy's algorithm, calculate $\sqrt{3,000}$ with a base-10 integer part and with a two-position base-60 fractional part. Then convert the answer to base-10.

FUN QUESTION 6.5: Calculate $\sqrt{3,000}$ to six-digit precision using the Babylonian squaring-the-rectangle method.

Chapter 7

PI (π)

In this chapter, we shall observe the evolution of algorithms to calculate the circumference and area of a circle, from the metric geometric algebra of the OB era, to Greek nonmetric geometric algebra, and finally to modern methods.

The symbol π is one of the most important constants in modern mathematics. It is defined by the relationship between the diameter and the circumference of a circle:

$c = \pi d$, where c = circumference of circle, d = diameter of circle   (7.1a)

This expresses the understanding that the ratio c/d is the same for all circles and is equal to π, a universal constant. When a mathematical relationship is intuitively obvious, it becomes a definition, so equation (7.1a) is essentially a definition.

The numerical value of π in equation (7.1a) can be determined by simply measuring the length of a string wrapped around a cylinder, or by measuring how far a wheel advances in one revolution. Such appreciation that $c \cong 3d$ probably goes back to the time of the invention of the wheel in the fourth millennium BCE in both Egypt and Mesopotamia. Even such simple measurements would have shown that $c > 3d$, but my guess is that the more complicated arithmetic required by a more precise value simply did not justify its use and what amounted to $\pi = 3$ was generally used.

On the other hand, the equation for the area of a circle must be derived and the same π appears in the correct derivation of the area of a circle,

$$A = \pi \left(\frac{d}{2} \right)^2 , \text{ where } A = \text{area of circle} \qquad (7.1b)$$

OB scribes did not usually use the concept of diameter but rather thought of the area in terms of the circumference, so the relationship of more relevance to them was

$$A = \frac{c^2}{4\pi}$$
(7.1c)

Such use of the π symbol is of modern origin. It was generally adopted following its use by the prestigious mathematician Leonhard Euler in 1736. When we equate ancient calculations of circumference and area of a circle to equations (7.1), we obtain what would have been the value of π if they had used equations (7.1). Comparing these values with the modern value, $\pi = 3.141592654$ (this is the precision exhibited by my electronic calculator), will be our measure of the precision of the ancient calculations.

RMP PROBLEMS 48 AND 50

RMP Problems 48 and 50 give the Egyptian derivation of the area of a circle.[1] This was important to know in order to calculate the capacity of cylindrical granaries (height $\times$ circle area). Figure 7.1 is a copy of the diagram and hieratic text of *RMP* 48. Figure 7.2 is the interpretation of the *RMP* 48 and 50 diagram and calculations.

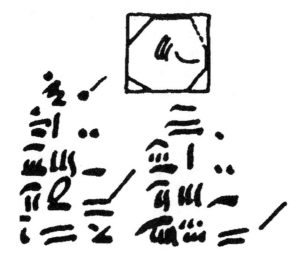

Figure 7.1 *RMP* Problem 48

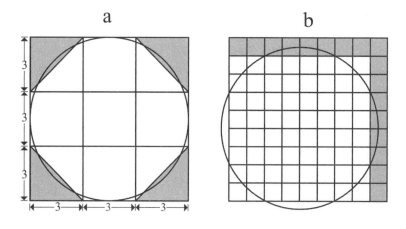

a   b

Figure 7.2 Egyptian derivation of area of a circle

Figure 7.2a illustrates how the *RMP* diagram approximates a circle by an octagon formed from a square with its corners cut off. The calculations are for a circle of diameter $d = 9$. The complete square has an area of $9^2 = 81$, while the cut-off corners have an area of 18 so that the approximated circle has an area of $81 - 18 = 63$. However, for the calculations, 64 rather than 63 was used, probably because the difference was not considered significant and $64 = 8^2$ produced an equation that was much simpler to use. Since the area of the approximated circle is 64, equation (7.1b) gives $64 = \pi(9/2)^2$, which yields $\pi = (16/9)^2 = 3.16$, a good approximation to 3.14, the modern value rounded off to the same number of digits. The Egyptian equation for the area of a circle was thus conveniently

$$A(\text{circle}) = (8d/9)^2 \tag{7.2}$$

rather than $A = 63(d/9)^2$. The choice of 64 rather than 63 gives a slightly more precise result, so the Egyptian result is fortuitously slightly better than the visualization warrants.

No surviving Egyptian document unambiguously calculates the circumference of a circle. However, the previous discussion related to figure 1.6 implies that they understood $A \propto c^2$. Gillings[2] reviews questionable interpretations of *MMP* Problem 10, which conclude that the circumference was calculated using equation (7.1a) with the same $\pi = (16/9)^2 = 3.16$ approximation.

OB PROBLEM TEXT YBC 7302

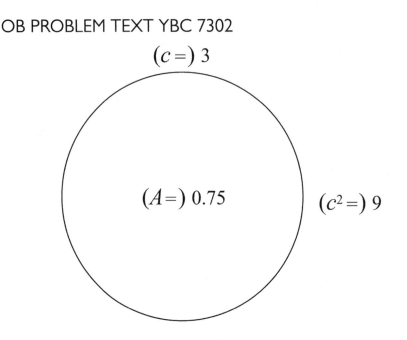

Figure 7.3 Neugebauer's interpretation of OB problem text YBC 7302

Figure 7.3 is a drawing of OB clay tablet YBC 7302. It is just a drawing of a circle with three numbers appended. I have transcribed these numbers into base-10 and added Neugebauer's[3] interpretation of the meaning of each number in parentheses. Using these numbers,

$$A(\text{circle}) = c^2/12 \qquad\qquad (7.3a)$$

and comparing this result to equation (7.1c) yields $\pi = 3$.

YBC 7302 gives no clue whatsoever about how the area of a circle was calculated. Figure 7.4 illustrates the simplest possible OB visualization that I can think of, which gives the area of circle as about 3/4 the area of a circumscribed square, so that

$$A(\text{circle}) = \frac{3}{4}d^2 \qquad\qquad (7.3b)$$

They also presumably knew by simple measurement that $c \cong 3d$, and using this relationship in equation (7.3b) yields equation (7.3a) as YBC 7302 requires. It is thus no coincidence that both equations (7.1a) and (7.1c) both yield $\pi = 3$. Both use the same OB approximation, $c \cong 3d$.

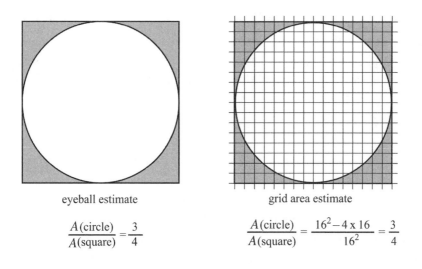

eyeball estimate	grid area estimate

$$\frac{A(\text{circle})}{A(\text{square})} = \frac{3}{4}$$

$$\frac{A(\text{circle})}{A(\text{square})} = \frac{16^2 - 4 \times 16}{16^2} = \frac{3}{4}$$

Figure 7.4 Possible OB visualization for derivation of area of a circle

A SCRIBE FROM SUSA CALCULATES π (ca. 2000 BCE)

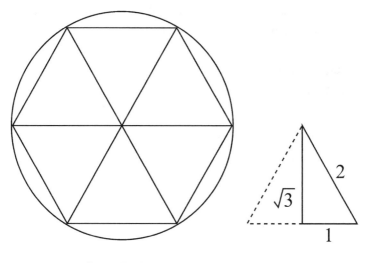

Figure 7.5 Hexagon inscribed in a circle

Neugebauer[4] thought that a fragment of a clay tablet was part of a calculation that approximated a circle by an inscribed hexagon. The tablet was identified as from Susa, capital of Elam, some 350 km east of Babylon in present-day

Iran. Elam's mathematics approximated that of Babylon. Figure 7.5 illustrates that the area of the hexagon is simply the area of six equilateral triangles. Using Neugebauer's notation, letting s_6 = side of equilateral triangle = side of inscribed hexagon, A_6 = area of the hexagon and referring to figure 7.5,

$$A_6 = 6\left(\frac{\sqrt{3}}{4}\right)s_6^2 \tag{7.4}$$

Neugebauer showed that using $\sqrt{3}$ = 1.75 that we found in chapter 6 was the

generally used OB approximation $6\left(\frac{\sqrt{3}}{4}\right) = 2.625$, which was exactly the value inscribed on the tablet.

Neugebauer interpretcd another entry of the tablet as relating the circumference of the circle, c, to the circumference of the hexagon, c_6

$$c_6 = 0.96c \tag{7.5}$$

Since $c = 2\pi s_6$ and $c_6 = 6s_6$, $c_6/c = 3/\pi$, and Neugebauer found that $3/\pi = 0.96$ if he assumed that the scribe had used a value of $\pi = 31/8$. Not only had his OB protagonists done a sophisticated calculation to determine the circumference of a circle, they also had a value for π that was every bit as precise as the Egyptian value.

Neugebauer's interpretation of equation (7.5) may be reasonable, but no explanation whatsoever of its derivation appears on the clay tablet, and not even Neugebauer proposed a derivation, so it is up to us to try.

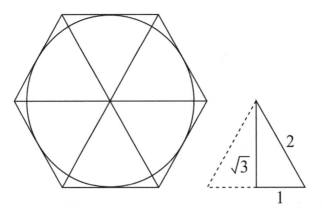

Figure 7.6 Hexagon circumscribed about a circle

When a hexagon is inscribed in a circle, as in figure 7.5, the area of the hexagon is obviously less than the area of the circle, and the circumference of the hexagon is obviously less than the circumference of the circle. However, when a hexagon is circumscribed about a circle, as in figure 7.6, then the area of the hexagon is obviously greater than the area of the circle, and the circumference of the hexagon is obviously greater than the circumference of the circle. Clearly, a better approximation for the area and circumference of a circle would be an average of the two visualizations.

Referring to figure 7.6 and adding to Neugebauer's notation so that now the side of the triangle is s_{6+}, the area of the hexagon is now A_{6+}, but the radius of the circle remains s_6, then $s_{6+} = \dfrac{2}{\sqrt{3}} s_6$ and therefore $A_{6+} = \left(\dfrac{2}{\sqrt{3}}\right)^2$.

Now letting the circumference and the area of the circle be given by the averages of the two hexagons, we obtain:

$$c = \frac{c_6 + c_{6+}}{2} = \frac{6}{2}\left(1 + \frac{2}{\sqrt{3}}\right)s_6 = 2\pi s_6 \quad \Rightarrow \quad \pi = \frac{3}{2}\left(1 + \frac{2}{\sqrt{3}}\right) \tag{7.6a}$$

$$A = \frac{A_6 + A_{6+}}{2} = \frac{6}{2}\left(\frac{\sqrt{3}}{4}\right)\left(1 + \frac{4}{3}\right)s_6^2 = \pi s_6^2 \quad \Rightarrow \quad \pi = \frac{7}{4}\sqrt{3} \tag{7.6b}$$

Using Neugebauer's determination that $\sqrt{3} = 1.75$ was the value generally used, we obtain that equation (7.6a) yields $\pi = 3.21$ and equation (7.6b) yields $\pi = 3.03$. The value of π derived from the circumference does not have to equal the value derived from the area because the definition of the circumference of a circle, equation (7.1a), was not used in the derivation of the area. What value for π did the scribe from Susa choose? He apparently did what OB scribes always did when faced with such dilemmas: he chose the average,

$$\pi = \frac{3.21 + 3.03}{2} = 3.12. \text{ We know this because of equation (7.5),}$$

$$\frac{c_6}{c} = \frac{3}{\pi} = \frac{3}{3.12} = 0.96$$

Thus, it appears that the scribe from Susa believed that the same value for π applied to both the definition of the circumference and the area of a circle. How did he know that? The first person to prove it was Archimedes, more than a millennium later. But a rigorous proof was not necessary in the OB era and YBC 7302 shows that one-π-for-all was already appreciated in Babylon, in Susa, and probably also in Egypt.

If we accept Neugebauer's interpretation that the scribe from Susa did indeed do the hexagon inscribed inside a circle, that he also did a hexagon circumscribed about a circle is reasonable. Archimedes, acknowledged as one of the greatest-ever mathematicians, considered the very same geometry to calculate π. It is a natural and intuitive visualization, consistent with the many other visualizations of divisions of regular geometric figures into triangles and other regular figures. It probably occurred independently to both the scribe from Susa and to Archimedes.

What was the object of the hexagon/circle calculation? Surely, the $\pi = 3$ approximation for area and circumference of a circle presented in YBC 7302 was convenient to use and was adequate for practical needs in a mud-brick civilization. Because the only way they knew to describe their mathematics was with metric geometric algebra, the conventional interpretation is that they were simply doing a practical calculation for a more precise answer. My guess is that, to the contrary, the scribe in Susa was doing the same thing that Greek mathematicians unambiguously did more than a millennium later with their nonmetric geometric algebra; he was satisfying his mathematical curiosity. His intuition told him that a hexagon was a better starting point for calculating properties of a circle than a square. He proved it and he was probably elated.

ARCHIMEDES CALCULATES π (ca. 200 BCE)

Metric geometric algebra was apparently taught to OB students by pointing to a sand-table sketch and writing or dictating the numerical calculation to be copied. Without this combination of oral and visual contact between teacher and student, transmission of geometric algebra was impossible in the OB era. Nor could the problems and their visualizations be very complicated with such limitations. The diagrams of figures 7.5 and 7.6 were adequate for this simple hexagon/circle visualization.

The Greeks solved the problem of requiring person-to-person transmission and also enabled consideration of more complex geometries by simply adding a few symbols (Greek letters) to the diagrams to define places. Since no copies of Greek mathematics prior to Euclid (323–283 BCE) survive, and Euclid used such annotated diagrams, we know that the concept was surely invented before about 300 BCE (see chapter 12). We can also appreciate why, without alphabetic writing, such a solution would have escaped OB and Egyptian consideration.

Figure 7.7 is the annotated diagram that I will use to present Archimedes's *Measurement of a Circle*, Proposition 2, "The ratio of the circumference of any circle to its diameter is less than $3\frac{10}{70}$ but greater than $3\frac{10}{71}$." Archimedes' original diagram and proof[5] is a bit complicated, so I am using a diagram and proof by Maor that is easier to understand but that adequately represents Archimedes' visualization/algorithm and the Greek annotated-diagram concept.[6]

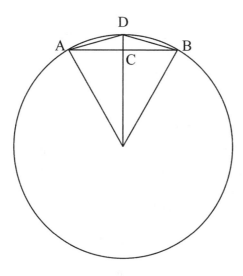

Figure 7.7 Polygon inscribed in a circle

Before beginning the proof, I would note that Archimedes' *Measurement of a Circle*, Proposition 1, is "The area of any circle is equal to a right-angled triangle in which one of the sides about the right angle is equal to the radius, and the other to the circumference of the circle." Algebraically expressed, this is simply $A = \frac{1}{2}ab = \frac{1}{2}r \times 2\pi r = \pi \left(\frac{d}{2}\right)^2$.

This is the first rigorous proof that the π in the definition of the circumference and the π in the calculation of the area are the same.

In a circle with a center at O and a radius OD, AB is one side of a polygon with n equal sides. OD bisects AB and hence $AD = BD$ are sides of a polygon with $2n$ equal sides. (Note how the annotations define lengths. Since the lengths are now algebraic symbols, they are printed in italic font, as is modern convention. The annotated letters on the diagram are place markers, not algebraic symbols, and are printed in normal font.) Such notation is a little more cumbersome than single-letter algebraic notation, but it is unambiguous and adequately simple. We want to find the side of the polygon with $2n$ sides, AD, in terms of the side of the polygon with n sides, AB. Applying the Pythagorean theorem to the triangle ACD (note how the annotations define geometric figures by letters at corners) and letting $OA = OD = 1$, we obtain

$$AD^2 = AC^2 + CD^2 = AC^2 + (OD - OC)^2 = AC^2 + (1 - OC)^2$$

Now applying the Pythagorean theorem to the triangle ACO, we obtain

$$OC = \sqrt{OA^2 - AC^2} = \sqrt{1 - AC^2}$$

Combining these two results, we obtain

$$AD^2 = AC^2 + \left(1 - \sqrt{1 - AC^2}\right)^2$$

To make understanding easier and to emphasize the advantage of modern notation over Greek notation, I will revert to modern, single-symbol algebraic notation by letting

$s_n = 2AC =$ side of an n-sided polygon

$s_{2n} = AD =$ side of a $2n$-sided polygon

which with some simple algebraic manipulation yields

$$s_{2n} = \sqrt{2 - \sqrt{4 - s_n^2}} \tag{7.7a}$$

Archimedes started his sequence of calculations with $s_6 = 1$. The beauty of equation (7.7a) is that starting with some easily calculated value for s_n, better approximations for the circumference are obtained by sequential use. The precision of this algorithm is limited only by the patience of the calculator. Archimedes' patience ran out at $2n = 96$. Appropriately, the algorithm is known as "the method of exhaustion."

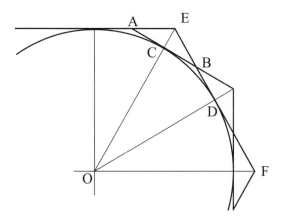

Figure 7.8 Polygon circumscribed about a circle

Next, Archimedes derived the circumference of a polygon circumscribed about a circle from the diagram presented in figure 7.8. For ease in drawing, the diagram is of a hexagon, but the proof is general. In terms of modern mathematical jargon, the diagram shows how a $2n$-gon is constructed from an n-gon. AB is the side of the $2n$-gon and EF is the side of the n-gon. Thus, we want to define AB in terms of EF. Again, we set the radius of the circle to unity, so $OC = OD = 1$. Now applying the Pythagorean theorem to the triangle EOD, we obtain

$$OD^2 + ED^2 = OE^2 = (OC + EC)^2 \Rightarrow 1 + ED^2 = (1 + EC)^2$$

Now we note that triangles EOD and ECB are similar because they have two equal internal angles. They both include a right angle and the angle CEB is common to both triangles. (Note how the annotations define angles: like for a triangle, three letters define an angle, but the middle letter must be at the apex.) Since the triangles are similar

$$EC/CB = ED/OD \Rightarrow EC = ED \times CB \Rightarrow 1 + ED^2 = (1 + ED \times CB)^2$$

Again reverting to modern algebraic notation, we obtain

$$s_{2n} = \frac{2\sqrt{4 + s_n^2} - 4}{s_n} \tag{7.7b}$$

Again, we start with a hexagon that the scribe from Susa gave as $s_{6+} = \dfrac{2}{\sqrt{3}}$ (see figure 7.6).

The circumference of an n-gon is simply $c_n = ns_n$. If this approximates the circumference of a circle $c = 2\pi r$ with $r = 1$, then

$$\pi = c_n/2 \tag{7.7c}$$

Equations (7.7) show that calculation of π first of all requires knowledge of $\sqrt{3}$ and then square roots of other numbers are also required. We have seen in chapter 6 that Archimedes' calculations of square roots were certainly clever, but not particularly good, especially his calculation of what he called the lower limit of a square root calculation. Thus, in Archimedes' Proposition 2 statement that $3\frac{10}{71} > \pi > 3\frac{10}{70}$, we do not know how much error is due to calculation of square roots and how much error is due to his n-gon algorithm. In order to test just the n-gon algorithm, figure 7.9 summarizes my calculations using "perfect" square roots, which I approximate by the ten-digit precision of my electronic calculator.

n	inscribed n-gon	circumscribed n-gon	average
6	3.0	3.5	3.25
12	3.1	3.2	3.15
24	3.13	3.16	3.145
48	3.139	3.146	3.1423
96	3.1410	3.1427	3.14185
Archimedes 96	3.140845 = 3 10/71	3.142857 = 3 10/70	3.14185
. . .			
3,072	3.1415921	3.1415937	3.14159293
. . .			
24,578	3.141592645	3.141592672	3.141592659
π			3.141592654

Figure 7.9 Calculations of π by Archimedes' algorithm assuming perfect square roots

For Archimedes' most-sided 96-gon, his calculation yields a value of π equal to the value calculated using perfect square roots. This means that all the imprecision in Archimedes' result is due to his n-gon algorithm. For the

96-gon, he calculated π to four-digit precision. Only for *n*-gons with many more sides is Archimedes' imprecision in square roots noticeable. My computer-program calculation shows that his algorithm requires a 3,072-gon to obtain seven-digit precision and a 24,578-gon to obtain nine-digit precision. Archimedes' algorithm is correct and beautiful, but it converges too slowly to give precise results in a pre-electronic-computer era. In our electronic-computer era, much better algorithms are now available.

FUN QUESTION 7.1: Unlike the scribe from Susa, Archimedes did not obtain a value for π by calculating the area of an *n*-gon. Can you?

The historical importance of Archimedes' algorithm is that it is a precursor to modern calculus that takes his "method of exhaustion" to the limit of an infinite number of subdivisions. But relevant to the scope of this book, Archimedes' derivation of his algorithm illustrates the huge importance of the Greek invention of annotated geometric diagrams. Without the annotations on figures 7.7 and 7.8, the ability to compose the algorithms and particularly to communicate the results other than person to person would be impossible. The difference between what the scribe from Susa did and what Archimedes' did does not lie either in the visualization or in the motivation, but in the annotation.

KEPLER CALCULATES THE AREA OF A CIRCLE (ca. 1600)

Johannes Kepler (1571–1630), one of the founders of modern astronomy, derived the area of a circle with the beautifully simple precalculus visualization of figure 7.10. As Archimedes had done, he approximated a circle as an inscribed *n*-gon. But he considered the area as a sum of triangles of base Δc. The Δ symbol has the meaning "a small piece of," so Δc is a small piece of the circumference. As the number of triangles approaches infinity, the height of each triangle approaches the radius of the circle, *R*, and the area of each triangle approaches

$$\Delta A = \frac{1}{2} R \Delta c$$

The total area of an infinite number of triangles is then

$$A = \sum \Delta A = \frac{1}{2} R \sum \Delta c = \pi R^2, \text{ because}$$

$$\sum \Delta C = 2 \pi R$$

The $\sum$ symbol has the meaning "summation of."

Note that the same π symbol appears in both the circumference and the area expressions because the circumference expression was used in the derivation of the area expression, just as it had been some three millennia earlier in OB clay tablet YBC 7302.

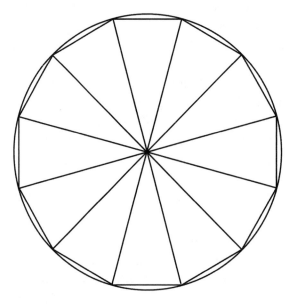

Figure 7.10 Kepler's derivation of area of a circle

EVERYBODY CALCULATES THE AREA OF A CIRCLE (ca. 2000)

Nowadays we also calculate the area of a circle by using the definition of the circumference, or more exactly, every student in Calculus 101 does, and there is no question that the same π appears in both the circumference and the area expressions. As shown in figure 7.11, the small piece of the circle summed can be an infinitesimally thin ring of circumference $2\pi r$ and thickness dr, so that

$$A = 2\pi \int_0^R r\,dr = \pi R^2$$

In calculus jargon and notation, summation is called integration, and the summation symbol, $\sum$, is replaced by the integration symbol, $\int$; the Δ symbol is replaced by the d symbol. Except for the fact that Kepler's derivation preceded the formal invention of calculus by Isaac Newton (1643–1727) and Gottfried von Leibniz (1646–1716), integration symbols could just as well have been used in his derivation.

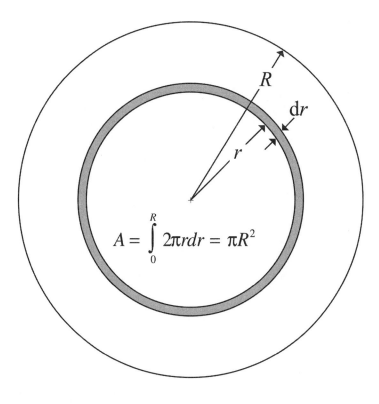

Figure 7.11 Calculus derivation of area of a circle

Thus from some four millennia ago to today, the visualization of dividing up a circle into little calculable pieces and summing them has not changed. Only the shape of the pieces, the notation, and the calculation technique have changed.

Chapter 8

SIMILAR TRIANGLES (PROPORTIONALITY)

I f all the internal angles of two triangles are equal, the triangles are similar. If a, b, c are the sides of one triangle and a', b', c' are the sides of a similar triangle, similar means that a/b = a'/b'and so forth. This is intuitively obvious, and both the Egyptians and the Babylonians understood and used similar-triangle relationships. In fact, anyone who attempts to draw a triangle similar to another easily observes that it is really only necessary to know two internal angles, so they certainly knew this also. The similar-figure concept is also frequently referred to as *proportionality*.

RMP PROBLEM 53

Figure 8.1a is a photograph of *RMP* Problem 53, an example of Egyptian use of similar triangles in a geometry problem. Figure 8.1b is Friberg's algebraic generalization and his interpretation of the drawing.[1] I have adopted Friberg's notation without any modification. The figure is reversed in order for the transcription of the data to read from left to right since the hieratic writing is from right to left. It is obvious that all of the triangles are similar because all have two angles in common, a right angle and a shared apex angle. The givens are A_1, A_2, d_2, and l_3. The task is to calculate A_3, d_1, s, l_2, and l_1. Friberg assumed a right triangle, although that is not obvious from the *RMP* drawing. However, drawings on both Egyptian and Babylonian mathematical documents seldom accurately represent the data and they only help define the problem. If it were not a right triangle, the problem would be unsolvable with the data given. It cannot be known whether the scribe

135

intended a right triangle or he just thought this a reasonable approximation for the actual triangle.

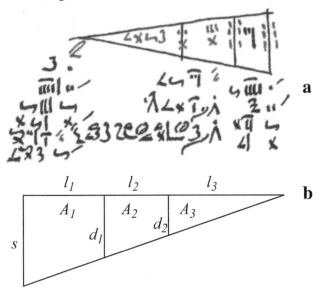

Figure 8.1 *RMP* Problem 53

a. A copy of the papyrus document b. Friberg's interpretation

I do not want to get involved in the tedious arithmetic of the problem, but I do want to consider how the teacher-scribe apparently composed the numerical values he assigned on his diagram. As was usually the case, he tried to assign simple numbers so that the students could concentrate on the geometry and not be distracted by complicated arithmetic. However, the complexity of this problem confused him and in his attempt to simplify things, he unintentionally complicated the arithmetic. The student who attempted to solve the problem simply quit after very few calculations.

Assignment of numerical values by the teacher had to be consistent with relationships between the three similar triangles

$$\frac{d_2}{l_3} = \frac{d_1}{l_2 + l_3} = \frac{s}{l_1 + l_2 + l_3}$$

His process of assigning numbers to the diagram was something like the following. First he chose $d_2 = 2$ and $l_3 = 6$, which gives a simple ratio $d_2/l_3 = 1/3$ and a simple area, $A_3 = 6$. Next he chose $l_2 = 9$, which required choosing

$d_1 = 5$ to maintain the same ratio, $d_1/(l_2 + l_3) = 1/3$. The area, $A_2 = 31 \ 1/2$, is again a simple number. Next he chose $l_3 = 8$, and to maintain the 1/3 ratio required that $s = 7 \ 2/3$, still a reasonably simple number, and for the calculation of the area he got $A_3 = 50 \ 2/3$. He then "simplified" the problem by changing A_3 to an integer, $A_3 = 50$, and making adjustments to make this consistent with the similar triangle requirements. That is when his numbers became confusing, the student quit, and that is where we shall leave arithmetic and consider the correct algebraic generalization.

Ignoring the actual numbers in the problem, the algebraic generalization of the solution to *RMP* Problem 53 is represented by the following equations:

- $r = [l_3/d_2]_{\text{given}} \Rightarrow r$
- $A_3 = d_2l_3/2 = rd_2^2/2 , \quad [d_2]_{\text{given}} \Rightarrow A_3$
- $A_2 + A_3 = rd_1^2/2, \quad [A_2]_{\text{given}} \Rightarrow d_1$
- $r = (l_2 + l_3)/d_1, \quad [l_3]_{\text{given}} \Rightarrow l_2$
- $A_1 + A_2 + A_3 = rs^2/2, \quad [A_1]_{\text{given}}, [A_2]_{\text{given}} \Rightarrow s$
- $r = (l_1 + l_2 + l_3)/s, \quad [l_3]_{\text{given}} \Rightarrow l_1$

A mathematics instructor presumably contrived the problem to teach student scribes how to use proportionality. However, the dimensions given in the papyrus are realistic numbers for farm-field sizes and the lesson may actually relate to the practical problem of parceling a triangular field. Although there are many variables involved, it is just a problem in sequential, elementary arithmetic, but we can certainly marvel at how, without algebraic symbols, it was possible to keep track of the many variables and their relationships. By positioning the values on the diagram, the student was expected to be able to visualize the relationships between variables.

OB PROBLEM TEXT MLC 1950

Neugebauer[2] translates and interprets a very similar OB problem text, MLC 1950, illustrated in figure 8.2. Figure 8.2a shows how the numbers were positioned on the clay-tablet drawing. The positioning convention for numbers on diagrams is that areas are inside figures and lengths are outside figures. We saw this convention used previously in figure 7.3 for OB problem text YBC 7302, which related to a circle. It is also the convention used in figure 8.1a for *RMP* 53. This is good evidence that Egyptian and OB mathematics interacted strongly, but who copied from whom is difficult to say definitively. OB mathematics appears to be somewhat more developed and hence Egypt was probably the copier.

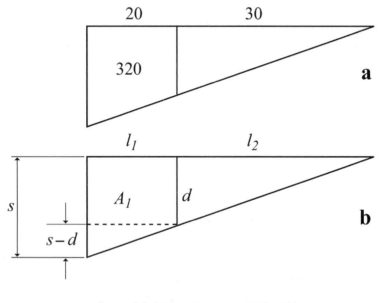

Figure 8.2 OB problem text MLC 1950

In figure 8.2, rather than using Neugebauer's notation, I have more or less adopted Friberg's as used for *RMP* 53 for easier comparison. The givens are A_1, l_1, and l_2. The problem is to find s and d. However, this problem requires solution of a pair of simultaneous linear equations. First, I will give the way I or anybody who uses modern symbolic algebra would easily solve this problem, and then I will give Neugebauer's interpretation of the convoluted solution used by the OB student to solve it.

The area of the trapezoid was well known as

$$A_1 = l_1 \frac{s+d}{2} \Rightarrow s+d = \frac{2A_1}{l_1} = \frac{2 \times 320}{20} = 32 \tag{8.1a}$$

A similar triangle relationship yields

$$\frac{d}{l_2} = \frac{s-d}{l_1} \Rightarrow d = \frac{l_2}{l_1+l_2}s = \frac{30}{20+30}s = \frac{3}{5}s \tag{8.1b}$$

Solving the pair of simultaneous equations

$$d+s = 32 \text{ and } d = \frac{3}{5}s, \text{ we get } s = 20, \; d = 12$$

Now let us consider the way the OB student-scribe solved this problem. The trapezoid area is treated almost as previously, obtaining

$$\frac{s+d}{2} = 16 \tag{8.2a}$$

Equation (8.1b) is treated quite differently, as

$$\frac{d}{l_2} = \frac{s-d}{l_1} \Rightarrow d = \frac{l_2}{l_1}(s-d) = \frac{3}{2}(s-d) \tag{8.2b}$$

Another similar triangle relationship is introduced

$$\frac{s}{l_1+l_2} = \frac{s-d}{l_1} \Rightarrow s = \frac{l_1+l_2}{l_1}(s-d) = \frac{5}{2}(s-d) \tag{8.2c}$$

Adding equations (8.2b) and (8.2c), we obtain

$$\frac{s+d}{2} = 2(s-d) \Rightarrow \frac{s-d}{2} = 4 \tag{8.2d}$$

Adding and subtracting equations (8.2a) and (8.2d), we obtain

$$\frac{s+d}{2} + \frac{s-d}{2} = s = 16+4 = 20 \tag{8.2e}$$

$$\frac{s+d}{2} - \frac{s-d}{2} = d = 16-4 = 12 \tag{8.2f}$$

Apparently, the only pair of simultaneous linear equations that this OB student knew how to solve was a pair with the form of equations (8.2e) and (8.2f). Recall that the solution of such a pair of this form was the way they solved all of the quadratic algebra problem texts of chapter 4.

OB PROBLEM TEXT IM 55357

OB problem text IM 55357 presents a different way of constructing similar triangles. It also presents an interesting example of the importance of notation in solving problems and it exhibits some intriguing properties with far-reaching implications. The tablet contains a diagram like that of figure 8.3, although roughly drawn on the original clay tablet. Here it is drawn exactly as it should appear if all the triangles were similar (3, 4, 5) Pythagorean triples, which is the conclusion of Hoyrup's interpretation of the diagram and his translation of the text.[3] A similar diagram would have been obtained with any choice of right triangle; the (3, 4, 5) triple was clearly chosen to make the arithmetic as simple as possible.

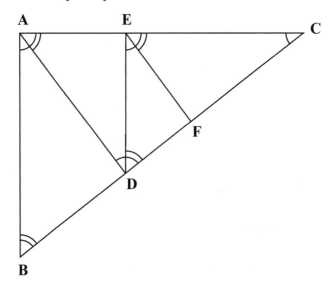

Figure 8.3 Diagram for OB problem text IM 55357

To make referring to the diagram unambiguous, I have added anachronistic Greek, alphabetic annotations to the diagram, as introduced in the discussion of Archimedes' calculation of π in chapter 7. Starting with the right triangle ABC, the line *AD* is drawn perpendicular to *BC*, then the line *DE* is drawn perpendicular to *AC*, and finally the line *EF* is also drawn perpendicular to *BC*. It is not as obvious as in the Egyptian example of figure 8.1 that this construction produces only similar triangles. How did the teacher who contrived this problem know that? Presumably, he used essentially the same reasoning as in *RMP* Problem 53.

Triangles ABC and ABD are similar because they have two angles in common, a right angle and the angle *ABC*. Note the new use of Greek notation to define angles with three letters, with the middle letter denoting the apex of the angle. In addition, triangles ABC and ADC are similar because they have two angles in common, a right angle and the angle *ACB*. However, if the triangles are similar, all internal angles are equal, thus the angle *BAD* in triangle ABD must be equal to angle *ACB* in triangle ADC. With such reasoning, all of the angles in figure 8.3 can be identified.

If you find this use of Greek annotation to define angles confusing, you are just one of many, which is why although it is theoretically unambiguous, it makes the solution of a somewhat complicated problem such as IM 55357 appear very difficult. Greek alphabetic annotation has the advantage of minimally encumbering of the diagram, and that is important in complicated drawings, but particularly when required to define angles it is prone to errors in both reading and writing. It is thus surprising that it was not until 1591 that essentially modern algebraic notation with single-letter symbols for quantities and operations was introduced by François Viete. One explanation for Greek retention of this cumbersome notation is that Greek mathematicians probably did not actually derive their proofs using the three-letter notation for angles that appears in their publications. My intuition is that such notation was just the convention for unambiguous written communication; for their own use, they must have used some less confusing method of defining angles. For example, a nonintrusive notation that is frequently used today and that I have used to define equal angles on figure 8.3 is that right angles are unmarked; angles that are equal are marked with the same number of arcs.

OB scribes, without any symbolic notation, nonetheless were able to reason that all the triangles in figure 8.3 were similar. In the OB era, person-to-person demonstration that all of the triangles in figure 8.3 are similar would have been easy, but written communication to someone not able to witness this being pointed out on a diagram would have been difficult if not impossible. It is true that modern translator/mathematicians have been able to understand the ancient documents but only with the advantage of having clues from subsequent Greek work, and even then not easily. By themselves, the OB era documents were not an efficient way to communicate non-person-to-person, nor were they intended for that. The Greeks were thus able to publish (if the making of handwritten copies can be called publishing) their mathematical inventions and the documents were saved in institutions such as the Library at Alexandria, where they were available for future generations of mathematicians to unambiguously, but not necessarily easily, understand and to build on. Mathematics in Greece progressed; mathematics in Egypt and Mesopotamia stagnated once it reached a level beyond which nonannotated visualization could not cope.

The reasoning that allowed OB scribes to recognize such division of a right triangle into similar triangles is identical to the rigorous proof in Euclid's *The Elements*, book VI, proposition 8, known to the cognoscenti and henceforth to us as simply Euclid VI-8. "If in a right triangle a perpendicular be drawn from the right angle to the base, the triangles adjoining the perpendicular are similar both to the whole and to one another." OB documents have generally been recognized as just calculations, although my guess is that some were probably much more. I think that IM 55537 is one of the much more.

There is an added insight that this construction produces. In figure 8.3, consider the right angle at A, angle *BAD* + angle *DAE* = right angle. From this result, it then follows that for every right triangle the sum of the internal angles is two right triangles.

FUN QUESTION 8.1: Use the Babylonian result, *for every right triangle the sum of the internal angles is two right angles*, to prove that *the sum of the internal angles of any triangle is equal to two right triangles*, which is just Euclid I-32. Euclid's proof does not use this method, but this method is just as valid and just as rigorous as Euclid's. We will probably never know whether anyone in Old Babylon ever came to this conclusion, but someone certainly could have. However, I doubt that the Babylonians ever arrived at this more general result because the problem would never have come up; they were not able to deal numerically with nonright triangles.

There is another even more surprising mathematical consequence of the IM 55457 method of dividing a right triangle into similar triangles. Consider the drawing of triangle ABC in figure 8.3. While the addition of Greek annotation to the figure represents a revolutionary advance in mathematical technique over Babylonian and Egyptian numerical calculation, modern algebraic notation is much easier to understand and we shall now also make another anachronistic transition to it by setting $AC = a$, $AB = b$, and $BC = c$.

Let us now define two ratios, $r_1 = a/c$ and $r_2 = b/c$. Now we can write the area of triangle ABC as

$$area \ (ABC) = \frac{1}{2} ab = const \times c^2 \propto c^2 \tag{8.3a}$$

where $const = (a/c)(b/c)/2$ and it is the same for all of the similar triangles in figure 8.3. Equation (8.3a) requires that the area of every similar triangle in figure 8.3 is proportional to the square of its diagonal (in modern jargon, its hypotenuse). Since obviously

$$area(ABC) = area(ADC) + area(ABD) \tag{8.3b}$$

$$c^2 = a^2 + b^2, \text{ the Pythagorean theorem!} \tag{8.3c}$$

This is essentially the Euclid VI-31 proof of the Pythagorean theorem. Were the Babylonians aware of this rigorous proof? We will probably never know. Note that the scribe who composed IM 55357 knew that it satisfied the Pythagorean theorem because the problem was defined by the (3, 4, 5) triple, but the proof of equations (8.3) is more general and does not depend on the fact that a (3, 4, 5) triple was used.

In chapter 9, we shall consider another interesting property of the figure 8.3 method of constructing similar triangles.

Chapter 9

SEQUENCES AND SERIES

ARITHMETIC SEQUENCES AND SERIES

In an *arithmetic sequence*, the terms have a *common difference* between them. The sequence of counting numbers, 1, 2, 3 ..., is certainly the most familiar example, with a common difference of one. Algebraically expressed, the basic relationship defining an arithmetic sequence of N terms, $a_1, a_2 \dots a_n \dots a_N$, with a common difference, d, is,

$$a_n - a_{n-1} = d \qquad (9.1a)$$

OB PROBLEM TEXT YBC 4608 #5

OB documentation that perhaps demonstrates understanding of arithmetic sequences is YBC 4608 #5, translated by Neugebauer:[1] "Six brothers inherit a triangle of land that is divided by equidistant lines parallel to the base." What is of particular interest here is that a problem generally considered purely algebraic is solved by a geometric visualization. Alternatively, it may just be another natural and intuitive way of defining a geometric problem and it just happens to represent an arithmetic series, of which the OB scribes were unaware. From OB problem texts it is impossible to know which interpretation is correct, but Egyptian texts appear to show awareness that they were indeed treating an arithmetic sequence and apparently derived their algorithms from the same visualization used by OB scribes. As usual, who copied from whom is an unanswered question.

Figure 9.1 presents the diagram according to Neugebauer's interpretation of the text. The *givens* in YBC 4608 #5 are the total area of the triangle, $A = 11.22{:}30_{60} = 11\frac{3}{8}$, and the length of the base, $l = 6.30_{60} = 6\frac{1}{2}$, so that the base of the triangle can be calculated from $A = \frac{la_1}{2}$, yielding $a_1 = 3\frac{1}{2}$, and from the figure, clearly, $d = -\frac{a_1}{6} = -\frac{7}{12}$. OB scribes preferred to start at the baseside of the triangle, making a_1 the longest base, and therefore d in equation (9.1a) is negative. The calculation actually carried out is simply repeated use of the basic relationship, equation (9.1a), to calculate all of the a_n's:

$$a_1 = 3\frac{1}{2}$$

$$a_2 = 3\frac{1}{2} - \frac{7}{12} = 2\frac{11}{12}$$

$$a_3 = 2\frac{11}{12} - \frac{7}{12} = 2\frac{4}{12}$$

$$a_4 = 2\frac{4}{12} - \frac{7}{12} = 1\frac{9}{12}$$

$$a_5 = 1\frac{9}{12} - \frac{7}{12} = 1\frac{2}{12}$$

$$a_6 = 1\frac{2}{12} - \frac{7}{12} = \frac{7}{12}$$

The presumed objective of the problem, to calculate the area inherited by each brother, was either not done or has been lost. It is easy to show that the areas, A_n, also form an arithmetic sequence. Each area is a trapezoid of height $l/6$ and from figure 9.1 it is easy to see that $A_n - A_{n-1} = -\frac{ld}{6}$. Since there is a common difference between sequential areas, the areas also form an arithmetic sequence.

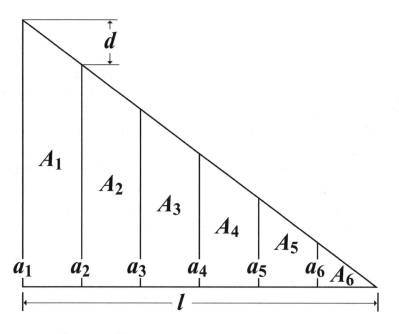

Figure 9.1 OB visualization of an arithmetic sequence

RMP PROBLEM 64

An arithmetic series is the sum of the terms in an arithmetic sequence. The sum of a series of N terms is thus

$$S_N = a_1 + a_2 + \ldots + a_n + \ldots + a_N \qquad (9.1\mathrm{b})$$

Now if we start with the last term, a_N, and apply the basic relationship of equation (9.1a), we can replace it with $a_{N-1} + d$. Continuing this process, essentially the series of operations applied to calculate all the a_n's in OB problem text YBC 4608 #5, we finally obtain

$$S_N = \frac{N}{2}\left[2a_1 + (N-1)d\right] \qquad (9.1\mathrm{c})$$

Alternatively, it can be visualized from figure 9.1 that the average term, $\langle a \rangle$, is clearly just the average of the first and the last terms and thus

$$S_N = N\langle a \rangle = N\frac{a_1 + a_N}{2} \tag{9.1d}$$

which is also how (9.1c) can be derived and interpreted. The visualization of figure 9.1 defines both the sequence and the sum of the series needed to solve Egyptian arithmetic sequence problems.

RMP Problem 64 reads "Divide 10 units of barley among 10 men so that the difference between each man and his neighbor is 1/8 unit."[2] Thus $S_N = 10$ units of barley, N = number of men (terms in the series) = 10, d = difference between sequential terms = 1/8. Gillings interprets the text as a literal application of equation (9.1c), which in practice meant writing an algorithm derived from the visualization of a diagram like figure 9.1,

$$S_N = \frac{N}{2}\left[2a_1 + (N-1)d\right] \quad \Rightarrow \quad 10 = 5(2a_1 + 9/8) \quad \Rightarrow \quad a_1 = 7/16 \tag{9.1e}$$

and the remainder of the a_n's are obtained by repeated application of basic equation (9.1a) of arithmetic sequences, $a_2 = 7/16 + 2/16 = 9/16 \ldots a_{10} = 25/16$.

RMP PROBLEM 40

The previous arithmetic sequence problems were rather trivial, but this problem is much more imaginative, to the point of being humorous. *RMP* Problem 40 reads "Divide 100 loaves among 5 men in an arithmetic sequence such that the total of shares to the 3 highest is 7 times the total of the shares to the 2 lowest."[3]

Again the scribe used the visualization of equation (9.1c) to account for the distribution of 100 loaves to the 5 men

$$100 = (5/2)(2a_1 + 4d) \tag{9.1f}$$

The distribution between the highest 3 and the lowest 2 is

$$(a_3 + a_4 + a_5) = 7(a_1 + a_2) \tag{9.1g}$$

Now applying equation (9.1a), this relationship was reduced to

$$3a_1 + 9d = 7(2a_1 + d) \tag{9.1h}$$

What we would do nowadays in this case would be to solve equation (9.1h) to yield $d = \dfrac{11}{2} a_1$, which we would then substitute back into equation (9.1f) to yield $a_1 = 100/60$, which would in turn yield $d = 55/6$. Now repeatedly using basic equation (9.1a), we would calculate all of the a_n's:

$$a_1 + \frac{10}{6}, \; a_2 = \frac{10}{6} + \frac{55}{6} = \frac{65}{6} \; \ldots \; a_5 = \frac{175}{6} + \frac{55}{6} = \frac{230}{6}. \text{ A bit tedious, but simple.}$$

The part of *RMP* containing Problem 40 is damaged, so interpretation requires some guessing and it is not surprising that there have been some differences of opinion. Gillings proposed a trial-and-error procedure, but a newer algebraic interpretation by Friberg appears more reasonable to me.[4] The Egyptian solution appears to be: Assume that $a_1 = 1$ and solve equation (9.1h) for d, which yields $d = \dfrac{11}{2}$. Now insert these values, $a_1 = 1$ and $d = \dfrac{11}{2}$, into the RHS of equation (9.1f) and obtain $(5/2)(2a_1 + 4d) = 60$ rather than the required 100. Therefore, it is necessary to multiply each a_n by 100/60 in order to obtain the correct distribution of loaves.

We have seen Egyptian use of this "method of false proposition" before to solve pairs of simultaneous equations in chapter 5, Berlin papyrus 6610 #1. OB scribes also used it and I give an example of such use in chapter 10. It is not only similarity in choice of problem that is evidence for interaction between Egypt and Babylon; it is also method of solution, the algorithm.

GEOMETRIC SEQUENCES AND SERIES

In a geometric sequence, the basic relationship defines sequential terms by a common ratio

$$\frac{a_n}{a_{n-1}} = r \tag{9.2a}$$

so a geometric series of N term is

$$S_N = a + ar + ar^2 + \ldots + ar^{n-1} + \ldots + ar^{N-1} \tag{9.2b}$$

Geometric sequences occur more in practice than arithmetic series and describe a number of important phenomena from population explosion to radioactive decay, but probably its most common application is in calculation of increase in savings (or debt) under conditions of compound interest. If your principal on day one is P_0 and you are receiving interest of p percent per year, at the end of a year your principal will be $P_0(1 + p/100) = P_0r$, so at the end of n years your principle will be $P = P_0r^n$.

There are many examples of Egyptian and OB geometric sequences and series. Friberg[5] summarizes OB table texts that exhibit geometric sequences, noting that it was common practice to list sequences up to r^{10}. He notes sequences deciphered by Neugebauer in the 1930s ($r = 9$ and 1 2/3) and more recently deciphered sequences ($r = 16$, 60/16, 1 1/6 …). He notes that the sequence for $r = 1$ 1/6 was also treated as a series and summed. There were no obvious practical applications of any of these tables, except perhaps to calculate compound interest, which based on these calculations ranged from generous to usurious. It is impossible to know whether these were conscious calculations of geometric series or just exercises in multiplication and addition.

By far the most quoted of all ancient mathematical documents is the geometric sequence and series calculated in *RMP* 79, which simply appears as a six-item list:

House	7 ($= a = r$)
Cats	49 ($= ar$)
Mice	343 ($= ar^2$)
Ears of Wheat	2,401 ($= ar^3$)
Quantity	16,807 ($= ar^4$)
Estate	19,607 ($= S_5$)

The reason for the frequent quotation is that the calculation has apparently survived and evolved into a Mother Goose rhyme enjoyed by children some four millennia later:

> As I was going to St. Ives,
> I met a man with seven wives.
> Every wife had seven sacks.
> Every sack had seven cats.
> Every cat had seven kits,
> Kits, cats, sacks, and wives,
> How many were going to St. Ives?

OB PROBLEM TEXT IM 55357—REVISITED

In chapter 8 we saw an insightful OB division of a right triangle into similar triangles in figure 8.3. Figure 9.2 continues this process of division indefinitely, and the areas of the similar triangles so produced form an infinite geometric series.

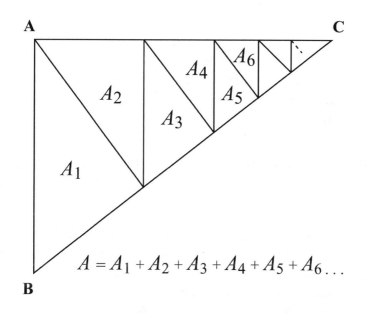

$$A = A_1 + A_2 + A_3 + A_4 + A_5 + A_6 \ldots$$

Figure 9.2 OB visualization of a geometric sequence

To see this, note that the hypotenuse, C_{i+1}, of each area, A_{i+1}, is just the side, a_i, of the preceding area, A_i. By equation (8.3a)

$$A_i \propto c_i^2 \text{ and } A_{i+1} \propto c_{i+1}^2, \text{ therefore } \frac{A_{i+1}}{A_i} = \left(\frac{a_i}{c_i} \right)^2 = r$$

and thus the area of the triangle ABC is given by the geometric series

$$A_{\text{ABC}} = A_1 + A_2 + \ldots = \sum_{i=1}^{i=\infty} A_i = A_1(1 + r + r^2 + \ldots A_1 \sum_{n=0}^{n=\infty} r^n \qquad (9.2c)$$

From the data given in chapter 8, we know that

$$\frac{A_{ABC}}{A_1} = \left(\frac{C}{C_1} \right)^2 = \left(\frac{5}{3} \right)^2 = \sum_{n=0}^{n=\infty} \left(\frac{4}{5} \right)^n \qquad (9.2d)$$

The diagram of problem text IM 55357 has led to the evaluation of a sum of an infinite geometric series. To see if we have done everything correctly, we can take the textbook answer for the sum of an infinite geometric series

$$\sum_{n=0}^{n=\infty} r^n = \frac{1}{1-r} = \frac{1}{1-\left(\frac{4}{5}\right)^2} = \frac{5^2}{5^2-4^2} = \left(\frac{5}{3}\right)^2 \qquad (9.2e)$$

and we see that our answer is correct.[6]

FUN QUESTION 9.1: Let the triangle that is divided by an infinite sequence of right triangles, as in figure 9.2, be a 30°, 60°, 90° triangle (half of an equilateral triangle). Let the hypotenuse of triangle ABC be 2. What is the area of A_4?

FUN QUESTION 9.2: Prove the bases, a_i's, in figure 9.2 also form a geometric series. For the triangle defined in the previous FUN QUESTION, what is a_{10}?

The question that intrigues me about problem text IM 55357 is how deep the OB scribes' understanding went. Did they see the proofs about internal angles and of the Pythagorean theorem demonstrated in chapter 8 to follow from this construction? Did they realize that the construction was a visualization of a geometric series?

When I was in high school, I generally felt that my teacher did not know much more than what was in the textbook and that the questions he put to us reflected the limits of his understanding. However, when I went to college I knew that the instructor who taught basic calculus was a talented mathematician and that the questions he put to us certainly did not represent the limits of his understanding. To which class do the Egyptian and the OB scribes belong? My guess is that, as today, there were members in both classes. To which class does the scribe who contrived problem text IM 55357 belong?

The works of the great Greek mathematicians Archimedes and Ptolemy, to name just two whose works have already been quoted in this book, were not written to be textbooks. They indeed reflect the deepest understanding of the authors. Their choice of subjects was governed by wherever their curiosity led them. Surely, among the Egyptian and the OB scribes there were also some who thought deeply about wherever their curiosity led them. We have certainly already seen ample evidence of OB interest in subjects that cannot be explained by anything other than mathematical curiosity.

The OB scribe who contrived problem text IM 55357 knew that the three triangles A_1, A_2, and A_3 formed a sequence where ratios of areas of adjacent triangles were the same; that was the essence of the numerical solution. He would certainly have realized that continued generation of other areas A_4, A_5 . . . , would also be defined by the same ratio. Perhaps it even occurred to him that a sum of many such triangles would at least approximate the area of triangle ABC as illustrated in figure 8.3.

Thus, he probably had a geometric visualization for a sequence of numbers that was not defined by the same difference between adjacent terms—what we now call an arithmetic series—but was defined by the same ratio between adjacent terms—what we now call a geometric series. My guess is that he contrived the specific problem text IM 55357 from a more general visualization of a geometric series and that problem text IM 55357 does not define the limits of his understanding. Although, what the limits of his understanding were we shall never know.

Chapter 10

OLD BABYLONIAN ALGEBRA: SIMULTANEOUS LINEAR EQUATIONS

U p to this point, all OB algebra has been metric geometric algebra. OB scribes learned how elementary figures related to each other in filling two-dimensional space. Indeed, they had a respectable understanding of plane geometry. Figure 10.1 sums up almost all the relationships covered.

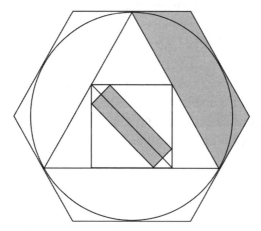

Figure 10.1 The ultimate OB geometric algebra problem

FUN QUESTION 10.1: In figure 10.1 a hexagon is circumscribed about a circle; an equilateral triangle is inscribed in the circle; a square is inscribed in the triangle; a rectangle is inscribed in the square. The rectangle (shaded) has an area $ab = 8$ and $a/b = 4$. What is the area of the trapezoid (shaded)? A good OB student could have solved this some four thousand years ago. Can you?

It is possible to solve this problem and all the OB problems considered so far with just a diagram to guide the calculation sequence without having to resort to symbolic algebra. In the text, I have anachronistically used algebraic symbols to make the descriptions of the operations short, unambiguous, and hopefully easier to understand. We have also seen that some OB calculations possibly imply rather rigorous proofs that we generally credit to Greek mathematicians of more than a millennium later. However, lack of OB use of generalizing symbolic notation or rhetorical commentary makes such intriguing implications just speculative.

I have criticized some of Neugebauer's OB interpretations as algebra when they were really geometric algebra, but I do not wish to imply that OB scribes never invented purely algebraic problems. They did and we shall consider one in this chapter. But to put the OB problem in perspective, let us first consider the simplest modern example of an algebra problem, the solution of a pair of linear simultaneous equations.

A MODERN ELEMENTARY ALGEBRA PROBLEM

At the beginning of all modern algebra texts are problems of the type: "The sum of the ages of Pat and Mike is 18. The difference between their ages is 5. Pat is older. How old are Pat and Mike?" The solution is to let x = Pat's age and y = Mike's age and to solve the pair of simultaneous linear equations with two unknowns:

$$x + y = 18 \qquad\qquad (10.1a)$$

$$x - y = 5 \qquad\qquad (10.1b)$$

By simply adding and then subtracting the two equations, we obtain that $2x = 23$ and $2y = 13$, then division by 2 yields $x = 11.5$ = Pat's age, and $y = 6.5$ = Mike's age.

In many OB "squares-and-rectangles problems" (see chapter 4), the last step in the solution is a pair of simultaneous linear equations with two unknowns:

$$(a + b)/2 = A \qquad\qquad (10.2a)$$

$$(a - b)2 = B \qquad\qquad (10.2b)$$

Again, simply adding and subtracting the equations yields $a = A + B$ and $b = A - B$. My guess is that OB scribes understood this simple algebraic operation without requiring any geometric visualization. But it is possible that it was visualized as illustrated in figure 10.2, and then once learned, perhaps visualization was no longer necessary. While the figure does geometrically solve equations (10.2), it is not an easy visualization and I doubt that it was used.

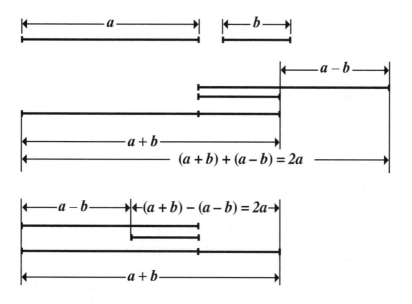

Figure 10.2 Geometric visualization of a pair of simultaneous linear equations

Equations (10.2) occur naturally in solutions to the OB "quadratic algebra" problems of chapter 4. As illustrated by OB problem text MLC 1950 in chapter 8, an OB scribe went to great lengths to convert the solution of a pair of simultaneous equations into the form of equations (10.2), presumably because that was the only way he knew how to solve algebraically a pair of simultaneous equations. In chapter 4, I had noted that Neugebauer had explained the solution of OB problem text VAT 8512 as an algebraic solution of three simultaneous equations. We can now see that this was attributing far too much algebraic prowess to OB scribes.

Now let us make the modern algebraic problem a little more complicated: "The sum of the ages of Pat and Mike is 18. The difference between
two-thirds of Pat's age and half of Mike's age is 5. Pat is the older. How
old are Pat and Mike?" Now the simultaneous equations are slightly more
complicated:

$$x + y = 18 \tag{10.3a}$$

$$2x/3 - y/2 = 5 \tag{10.3b}$$

There are several solution options. For example, with just some simple multiplications of both equations, we obtain the pair of equations:

$$3x + 3y = 54 \tag{10.4a}$$

$$4x - 3y = 30 \tag{10.4b}$$

Now addition of the two equations yields $7x = 84$, or $x = 12 = $ Pat's age. Substituting this value for x in any equation yields $3y = 18$, or $y = 6 = $ Mike's
age. This is also not difficult for most students today.

However, neither the ancient Egyptians nor the Babylonians were
capable of such an algebraic solution of even such a simple problem. My
guess is that the reason for this is that with algebraic notation, it is easy to
see what arithmetic is required to extract x and y individually, but with only
OB rhetorical algebra the necessary arithmetic is obscured, and without any
simple geometric construction available, a different algebraic solution was
required.

OB PROBLEM TEXT VAT 8389

To illustrate this point, let us now consider an OB problem text translated by
Neugebauer before 1935.[1] "The yield in volume units of grain per unit area
from one field is 2/3. The yield from a second field is 1/2. The sum of the
areas of the two fields is 1,800. The difference between yields of the two
fields is 500. What is the area of each field?" As usual for such problems
using modern algebraic notation, let x and y be the areas of the two fields and

thus it is necessary to solve the pair of simultaneous linear equations with two unknowns:

$$x + y = 1,800 \tag{10.5a}$$

$$2x/3 - y/2 = 500 \tag{10.5b}$$

This is essentially the same as the more complicated Pat and Mike problem just given, and essentially the same algebraic solution now yields for the areas of the two fields: $x = 1,200$ and $y = 600$.

The OB solution was first to assume the *false proposition* that both fields had equal areas of 900. If we substitute $x = y = 900$ in the equation (10.5b), we obtain that $2x/3 - y/2 = 150$ rather than 500, so to make up the deficit of $500 - 150 = 350$, it is clear that it is necessary to add area to the first field and subtract the same area from the second field. If one unit of area is transferred, the yield of the first field increases by 2/3 grain volume units and the yield of the second field decreases by 1/2 grain volume units so that the difference increases by $2/3 + 1/2 = 7/6$. Thus it is necessary to transfer $350/(7/6) = 300$ area units and so $x = 900 + 300 = 1,200$ and $y = 900 - 300 = 600$. Certainly, a clever algebraic solution, but just not what we do today.

We have previously seen the method of *false proposition* used in Egypt to solve Berlin Papyrus Problem 6610 #1 in chapter 5 and *RMP* Problem 40 in chapter 8. It is not only similarity in choice of problem that is evidence for interaction between Egypt and Babylon; it is also method of solution, the algorithm.

Chapter 11

PYRAMID VOLUME

In prior chapters we have considered how scribes in Egypt and Babylon conceived of filling two-dimensional space with simple two-dimensional figures (triangles, squares, rectangles …), the beginning of what we today refer to as plane geometry. Now we shall consider how they conceived of filling three-dimensional space with simple three-dimensional figures (cubes, prisms, pyramids ...), the beginning of what we today refer to as solid geometry.

Both Egyptian and OB scribes probably knew that the volume of a prism of height h and of base area A of any shape was given as

$$V = hA = ha^2 \text{ (prism with a square base)} \qquad (11.1a)$$

although records of calculations only survive for a few shapes of base. Egyptian *RMP* Problem 41 gives the volume of a cylindrical prism,[1] and OB problem text YBC 5037 gives the volume of prisms with square and rectangular bases.[2]

We shall shortly see calculations that indicate that both Egyptian and OB scribes knew the correct equation for the volume of a pyramid,

$$V(\text{pyramid}) = V(\text{prism})/3 = ha^2/3 \text{ (pyramid with a square base)} \quad (11.1b)$$

Square bases have been the shape of choice for pyramids throughout the ancient world, for Egyptians, Babylonians, Mayas, and probably many others, and that is the shape I shall primarily consider. Figure 11.1 illustrates these definitions.

Nowhere in this ancient record is there an unambiguous clue as to how Egyptians or Babylonians derived equation (11.1b) for pyramid volume. All suggestions of pre-Greek derivations of pyramid volume are thus conjecture, but they are interesting examples of the inventiveness of modern historians/mathematicians and may even have some relevance to how OB-era scribes thought. The exercise is to use documentary evidence and reasonable assumptions about what was known four thousand years ago to invent a probable derivation. It is such an intriguing problem that the number of proposed explanations of how the pyramid-volume equation was derived is only exceeded by the number of proposed explanations of the mystic properties of pyramids.[3]

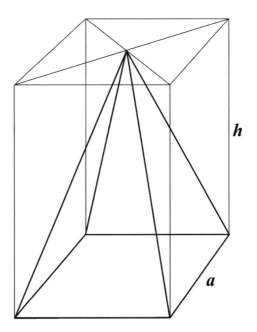

Figure 11.1 Prism and pyramid with a square base

HOW THEY KNEW V(pyramid)/V(prism)=1/3

From figure 11.1, V(pyramid) is clearly some fraction of V(prism), but that the fraction is 1/3 is not obvious. Figure 11.2 shows that if we subtract two wedges, easily visualized (at least by me), of combined volume of half the prism, that leaves a pyramid plus some undefined volume. Thus some ancient scribe could be sure that V(pyramid) < V(prism)/2 and a fraction of 1/3 was a reasonable guess, but whether or not correct, it may have been considered sufficiently precise.

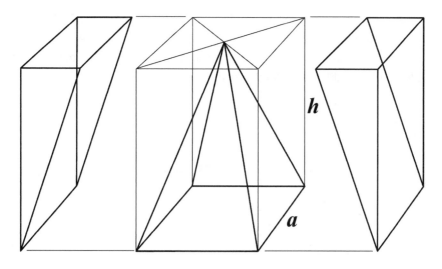

Figure 11.2 Proof that a pyramid is less than half a prism

Another reasonable conjecture is that someone could have simply compared the weights of a clay model of a pyramid and a clay model of a prism, of equal base and height, and thereby have empirically determined equation (11.1b).[4]

Figure 11.3 shows how a cube could have been visualized as sectioned into six pyramids, yielding V(pyramid) = V(prism)/3. This proof is only for the specific height-to-base ratio $h/a = 1/2$, but that could have been a sufficient clue to make the reasonable guess that the result would be the same, or at least a reasonable approximation, for any h/a. I find this conjecture, despite the fact that it yields the correct exact value for the fraction, a rather too sophisticated three-dimensional visualization for the OB era, but we cannot rule it out.

The method that I think could most likely have been used is that someone visualized or perhaps even built a model of a pyramid and simply counted the blocks, N(pyramid), and compared that number to the number of blocks in a prism, N(prism), of the same base and height. For modern structures made with poured concrete, volume is the correct measure of material quantity. For Egyptian pyramids made by stacking blocks of stone, counting blocks would have been the natural measure of material quantity. In the Egyptian pyramid-building era, ca. 2700–2400 BCE, surely many calculations requiring block counting must have been made to determine material and labor requirements, but no such documentation survives. In Babylon, where construction was primarily with mud bricks, surviving clay tablets show that calculations were routinely performed to define how many bricks would be obtained from a specific volume of mud.[5]

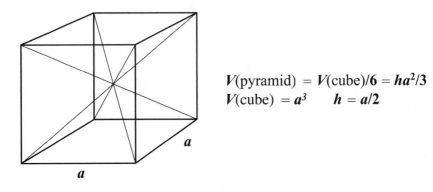

$$V(\text{pyramid}) = V(\text{cube})/6 = ha^2/3$$
$$V(\text{cube}) = a^3 \quad h = a/2$$

Figure 11.3 Sectioning of a cube—derivation of pyramid volume

As the number of blocks in the model increases, the ratio of $N(\text{pyramid})/N(\text{prism}) = V(\text{pyramid})/V(\text{prism})$ approaches the correct value. That would have been intuitively appreciated because we have already seen (see chapter 7) both Egyptian and OB use of division into small squares and triangles to approximate the area of a circle. Figure 11.4 illustrates an application of this conjecture with an easily constructed and counted sequence of layers: proceeding from top to bottom, a layer with $(N + 1)^2$ blocks follows every layer with N^2 blocks.

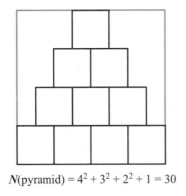

$N(\text{pyramid}) = 4^2 + 3^2 + 2^2 + 1 = 30$

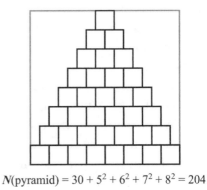

$N(\text{pyramid}) = 30 + 5^2 + 6^2 + 7^2 + 8^2 = 204$

Figure 11.4 Counting blocks method of determining pyramid volume

- For the four-layer pyramid: $N(\text{pyramid}) = 1 + 2^2 + 3^2 + 4^2 = 30$, and $N(\text{prism}) = 4^3 = 64$.
- For the eight-layer pyramid: $N(\text{pyramid}) = 30 + 5^2 + 6^2 + 7^2 + 8^2 = 204$, and $N(\text{prism}) = 8^3 = 512$.

If we do the calculation the way we normally do nowadays with an electronic calculator and express fractions decimally, we get:

- For the four-layer pyramid: $N(\text{pyramid})/N(\text{prism}) = 30/64 = 0.469$.
- For the eight-layer pyramid: $N(\text{pyramid})/N(\text{prism}) = 204/512 = 0.398$.

The ratio may be converging toward some limit as layers are added, but it would be a long and tedious calculation to determine the limiting ratio precisely. However, if we calculate this ratio in the manner that the Egyptians expressed fractions, the probable convergence to 1/3 becomes obvious. Applying the greedy algorithm as described in chapter 2 to calculate $N(\text{pyramid})/N(\text{prism})$ in terms of Egyptian fractions, we obtain:

- For the four-layer pyramid: $N(\text{pyramid})/N(\text{prism}) = 30/64 = 1/3 + 1/8 + 1/96$.
- For the eight-layer pyramid: $N(\text{pyramid})/N(\text{prism}) = 204/512 = 1/3 + 1/16 + 1/348$.

It is already quite clear that the ratio of $N(\text{pyramid})/N(\text{prism})$ will probably converge to the value of 1/3 for large values of N and hence that probably $V(\text{pyramid}) = V(\text{prism})/3$. It might be thought that this derivation applies only to one pyramid of height-to-base ratio $h/a = 1$. However, the blocks do not have to be cubes as happens to be the case in figure 11.4, but we can use squat prisms or tall prisms and hence the slope of the pyramid can be anything required.

FUN QUESTION 11.1: Apply the greedy algorithm (see chapter 2) to calculate $N(\text{pyramid})/N(\text{prism}) = 30/64$ and $N(\text{pyramid})/N(\text{prism}) = 204/512$ in terms of Egyptian fractions.

It is possible to extend this method of counting blocks to yield a general analytic solution and hence a rigorous proof, and not just numerical results that provide a strong hint of what V(pyramid) must be. The calculation, which I do in terms of possible OB competence, is rather tedious and therefore I present it in appendix B for those who are interested in such details. The result of this calculation is:

$$N(\text{pyramid}) = M(M+1)(2M+1)/6 \qquad\qquad (11.2)$$

where M^2 = number of blocks in the bottom layer. If M is a large number, such that $M \gg 1$, equation (11.2) approaches N(pyramid) = $M^3/3$ and hence N(pyramid) = N(prism)/3. What is intriguing about this result is that a calculation found on a cuneiform clay tablet goes a long way, if not quite all the way, toward a derivation of equation (11.2). This cuneiform tablet is also discussed in appendix B. I think that this calculation is too difficult to be a realistic proposal for the OB era, but I present it to illustrate that the solution given by Euclid to be presented next is not the only way that this problem could have been solved in a precalculus era.

EUCLID PROVES V(pyramid)$/V$(prism) = 1/3

The first surviving record of a rigorous proof of this result is Euclid XII-7. Euclid (ca. 300 BCE) was not the originator of all his proofs, nor does he identify the originators, but scholars have concluded that Euclid XII-7 is probably attributable to Eudoxus (ca. 410–350 BCE).

Euclid XII-7 epitomizes the difference between Greek and Egyptian/Babylonian geometric algebra. In chapter 7, we saw the importance of the Greek invention of adding annotations to their geometrical diagrams, simply Greek letters to mark locations. In Archimedes' use of such annotations, we noted their importance in unambiguous communication of results. In Euclid's use of Eudoxus's work we see that annotation enabled Euclid to understand results from a prior generation. The annotations possibly helped Archimedes develop his algorithm but were certainly not necessary. In Euclid XII-7, we shall see that the alphabetical annotations play a new and essential role in defining the logic of the proof. Such Greek annotation is not the only possible annotation that could have been used, but it does the job in a wonderfully elegant proof.

Euclidian proofs can be difficult and the three-dimensional visualization required in XII-7 can be troublesome. Thus, I shall not simply give an exact reading of Euclid but will modify the proof to make it easier to understand. I shall also introduce some new variations in use of the Greek annotations that are self-explanatory.

Consider the triangular pyramid, ABCD, in figure 11.5a. The figure has four sides and so it can also be called a tetrahedron. We can choose any of the four sides to be a base. When we choose ABC for the base and D for the apex, we use the notation D-ABC. It makes no difference which corner we choose for the apex. It is still the same pyramid with the same volume and thus:

$$V(\text{D-ABC}) = V(\text{A-BDC}) = V(\text{B-ACD}) = V(\text{C-ABD})$$

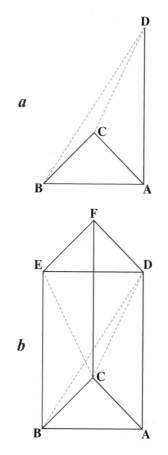

Figure 11.5 Diagrams relevant to Euclid XII-7

Figure 11.5b is of a prism with a triangular base, ABCDEF. The line *EC* bisects the rectangle BEFC, the line *BD* bisects the rectangle ABED, and the line *CD* bisects the rectangle ACFD, thereby dividing the prism into three triangular pyramids all with an apex at C. (You may have to stare at the diagram for a while for the perspective to come into focus.)

$$V(\text{prism}) = V(\text{C-ABD}) + V(\text{C-DEF}) + V(\text{C-BED}) \qquad (11.3)$$

The three pyramids are not all of the same shape, but XII-7 proves that all three of these triangular pyramids have the same volume.

- Since *BD* bisects the rectangle ABED into two equal triangles, ABD and BED, $V(\text{C-ABD}) = V(\text{C-BED})$.
- Since *EC* bisects the rectangle BEFC into two equal triangles, BEC and CEF, $V(\text{D-BEC}) = V(\text{D-CEF})$.
- Changing the apexes of both pyramids: $V(\text{D-BEC}) = V(\text{C-BED})$ and $V(\text{D-CEF}) = V(\text{C-DEF})$, thus we obtain $V(\text{C-DEF}) = V(\text{D-CEF}) = V(\text{D-BEC}) = V(\text{C-BED})$.

We have proved that in equation (11.3) both $V(\text{C-ABD})$ and $V(\text{C-DEF})$ equal $V(\text{C-BED})$ and so all three pyramids are of equal volume and hence we have proved that for triangular prisms, $V(\text{pyramid}) = V(\text{prism})/3$. Not an easy proof to follow, especially with the confusing Greek notation, but it is a quintessential expression of consequences of the transition from metric to nonmetric geometric algebra. What a difference the millennium made.

Two prisms with right-angle isosceles triangle bases can form a prism with a square base and so we have also proved that this also holds for a prism with a square base. In fact, triangular prisms can be stacked to form a base of any shape. A number of triangular prisms can even approximate a circle (see figure 7.10, Kepler's derivation of the area of a circle) and so it is possible to extend this proof to apply to conical prisms, which is in fact Euclid XII-10.

We have not exhausted the supply of conjectures about how, some four thousand years ago in Egypt and Babylon, they knew how to calculate the volume of a pyramid with a square base. What is certain is that they had a method that cannot be identified with any certainty, which strongly suggested that $V(\text{pyramid}) = V(\text{prism})/3$ was a good choice, and they chose it.

TRUNCATED PYRAMID (FRUSTUM) VOLUME

The Egyptian calculation of the correct volume of a frustum with a square base has been called the "zenith of Egyptian mathematics." Figure 11.6 illustrates that the frustum is just a pyramid with its top cut off and so the shape is also referred to as a truncated pyramid. The volume calculation is problem 14 in the *Moscow Mathematical Papyrus* (*MMP* to the cognoscenti and to us hereon). Apparently written in the eighteenth century BCE, XIII dynasty period, it is thus contemporary with OB mathematics. There is no precedent for such a sophisticated derivation in surviving Egyptian mathematical papyri, and some scholars even have proposed that it may have been just a lucky guess. Rather, as we shall see, it was not a lucky guess, but a very good conjecture based on sound reasoning, although understandably far short of the rigorous logic of Greek mathematicians of more than a millennium later.

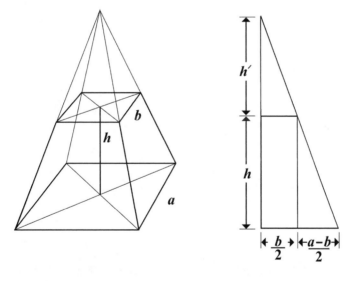

Figure 11.6 Frustum with a square base

Since 1927, when *MMP* 14 was first translated, it has been an irresistible attraction for conjectures about how this solution was derived. The problem with all of these conjectures is that there has been no way to establish which solutions are reasonable because of a total lack of any complementary Egyptian documentation. However, taking into account new translations of related OB calculations, we shall now be able to understand the evolution of this solution.

Unlike most mathematical documentation in Egypt and Babylon, a drawing fortunately accompanies *MMP* 14. Without this diagram, presented in figure 11.7, rough as it is, it would have been difficult to conclude unambiguously that the calculation was indeed that of the volume of a frustum with a square base. The notations on the diagram give the dimensions and the solution algorithm in hieratic script.

Figure 11.7 *MMP* Problem 14

The translation of *MMP* 14, essentially as given by Gillings,[6] expressed in *rhetorical algebra* that was used similarly in both Egypt and Babylon, and the algebraic generalization in Gillings's notation, is given in figure 11.8. The algebraic generalization of the algorithm shows that in effect the solution was of the equation

$$V(\text{frustum}) = h(a^2 + ab + b^2)/3 \tag{11.4a}$$

The scribe who in effect derived this equation could not have known that he had the correct equation. At best he could have known that in the limit of $b = 0$, he obtained $V(\text{pyramid}) = ha^2/3$, and in the limit $a = b$, he obtained $V(\text{prism}) = ha^2$, thus $V(\text{pyramid}) = V(\text{prism})/3$. He knew that he had the correct solution at these limits and it would have been reasonable to expect that he at least had a useful approximation in between these limits. I think he would have been surprised to learn that he had the exact solution at all values of a/b.

Step	Translation of *MMP* 14	Algebraic generalization
1	Truncated pyramid of height 6	h
2	Base of 4 and top of 2	a, b
3	Square 4, result 16	a^2
4	Double 4, result 8	ab
5	Square 2, result 4	b^2
6	Add 16 to 8 and 4, result 28	$a^2 + ab + b^2$
7	Calculate 1/3 of 6, result 2	$h/3$
8	Twice 28, result 56 is correct	$V = h(a^2 + ab + b^2)/3$

Figure 11.8 Translation and algebraic generalization of *MMP* Problem 14

It is important to emphasize that the *MMP* 14 algorithm is not a derivation of equation (11.4a). This algorithm is just the series of arithmetic steps that calculate the volume of a geometric visualization. The first thing we note about the translation is that although step 4 states that the operation is a doubling of the base, the algebraic generalization is given as the product of top and base. In addition, it is unclear that line 8 refers to the result of line 7, as assumed in the algebraic generalization. To make mathematical sense of the calculation, Gillings assumed that the Egyptian scribe made a simple copying error and used ambiguous language. Such corrections of apparent errors and assumptions about content in damaged or missing portions of documents are an unavoidable source of ambiguity about content. The conclusion that this calculation is indeed that of a truncated pyramid appears justified.

In order to establish that *MMP* 14 does indeed give the correct solution, let us rigorously calculate the frustum volume by calculus. Even if you do not understand calculus, you can accept the result knowing that every year literally thousands of students in Calculus 101 confirm it.

$$V(\text{frustum}) \int x^2 \, dy = \frac{h}{a-b} \int_b^a x^2 \, dx = \frac{h}{3(a-b)}(a^2 - b^3) \qquad (11.4b)$$

FUN QUESTION 11.2: Show that volume equation (11.4b), derived by calculus, is equivalent to the algebraic generalization of the *MMP* 14 calculation of equation (11.4a).

The calculus derivation is based on the geometric visualization of figure 11.9. It was not until millennia latter that Isaac Newton (1642–1727) and Gottfried von Leibniz (1646–1716) invented calculus that enabled calculation based on this visualization of a pyramid as a sum of an infinite number of infinitely thin plates. However, this visualization of a pyramid as a sum of square plates must have been obvious to Egyptian scribes who had only to look at their pyramids, which had been standing for about seven hundred years by the time that *MMP* was written.

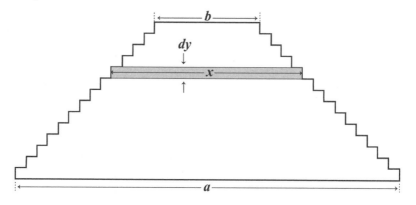

Figure 11.9 Visualization of frustum volume for calculus derivation

Now that we are reasonably confident that *MMP* 14 really is a calculation of frustum volume, let us consider some of the modern conjectures about how the Egyptians obtained this result. Unlike the calculus derivation of the frustum volume, all conjectured derivations rely on prior knowledge of the volume of a pyramid. The simplest visualization for calculating the volume of a frustum is to subtract the volume of the cut-off top prism from the volume of the whole pyramid. Referring to the notation of figure 11.6:

$$V(\text{frustum}) = [(h + h')a^2 - h'b^2]/3 \qquad (11.4c)$$

We can calculate the only variable in equation (11.4c) that is not one of the givens, h', by the similar triangle relationship illustrated in figure 11.6.

$$\frac{h'}{h} = \frac{a}{a-b} \qquad (11.4d)$$

In light of chapter 8, we can be confident that similar triangle relationships were understood. However, the numerical evaluation of h' by equation (11.4d) could not have been used in *MMP* 14. Such a calculation simply does not appear anywhere in the *MMP* 14 algorithm. The other possibility is that equation (11.4d) was used to eliminate the unknown h' in equation (11.4c). But if we do this, we do not get equation (11.4a), which is required by the *MMP* 14 algorithm. Rather, we get equation (11.4b), the calculus solution. We know from the answer to FUN QUESTION 11.2 that we can transform equation (11.4b) into equation (11.4a) as required to satisfy *MMP* 14, but only with a sophisticated algebraic factorization. It seems unlikely that this was a XIII dynasty or OB capability and we must look for a different visualization.

A consequence of visualizing a pyramid as a sum of plates is that although the pyramid of figure 11.9 is symmetric, we can visualize the individual prisms sliding horizontally without changing the volume. In mathematical jargon, a pure, shear transformation of any body does not alter its volume. If the jargon is confusing, simply visualize a neatly stacked deck of cards with an easily calculated volume. Now splay the cards out in any complex configuration with a volume that would be difficult to calculate; the volume remains constant at fifty-two times the volume per card. Thus, the volume of the pyramid of figure 11.10 with two perpendicular sides, which makes for an easier visualization and calculation, is the same as the volume of a symmetric pyramid.

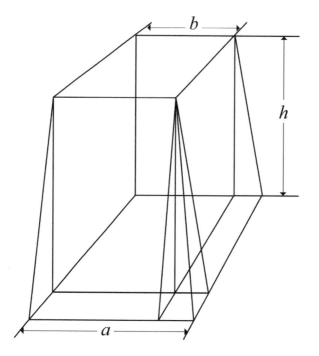

Figure 11.10 Visualization of frustum in volume derivation

The frustum can be visualized as the sum of calculable volumes of a pyramid plus two wedges plus a prism,

$$V(\text{frustum}) = h(a - b)^2/3 + hb(a - b) + hb^2 = h[(a - b)^2 + 3b(a - b) + 3b^2]/3 \quad (11.4e)$$

Equation (11.4a), which summarizes the *MMP* 14 algorithm, has three terms. Equation (11.4e) also has three terms and is just as easily calculated, but the calculation is not compatible with the *MMP* 14 algorithm. Even without comparing numerical values of each term, we see that there is no subtraction operation in the algorithm as required to evaluate equation (11.4e). Neugebauer believed that the Egyptians were capable of the algebraic transformation required to convert equation (11.4e) into equation (11.4a),[7] but Waerden was skeptical and proposed another visualization. I think it improbable that either Neugebauer's or Waerden's solution is correct, but that Waerden was at least on the right track in looking for a visualized solution since we now know that this was the basis for essentially all OB algebra rather than Neugebauer's purely algebraic interpretations.

FUN QUESTION 11.3: Show that equation (11.4e) can be algebraically converted into equation (11.4a).

However, new translations and interpretations of OB problem texts *BM* 85194 and *BM* 85196 by Friberg imply radically different derivations.[8] Two approximate frustum solutions that were used as acceptable approximations are

V(frustum) = $h[(a + b)/2]^2 = ha^2/4$,
in the limit $b = 0$ (frustum = pyramid)   (11.5a)

V(frustum) = $h(a^2 + b^2)/2 = ha^2/2$
in the limit $b = 0$ (frustum = pyramid)   (11.5b)

In other words, in the limit of a complete pyramid where the correct result should be $ha^2/3$, equation (11.5a) gives a result that is too low and equation (11.5b) gives a result that is too high. However, in the limit of $b = a$ both equations give the correct answer of ha^2 and so they are reasonable approximations for pyramids with steep slopes. However, in the same OB documents correct solutions are also given as:

V(frustum) = $h[(a + b)/2]^2 + h(a - b)^2/12$ (11.5c)

V(frustum) = $h(a^2 + b^2)/2 - h(a - b)^2/6$ (11.5d)

FUN QUESTION 11.4: Show algebraically that equations (11.5c and 11.5d) can be transformed into the *MMP* 14 solution, equation (11.4b).

Apparently, they arrived at the correct solutions by starting with the approximate solutions of equations (11.5a and 11.5b) and then sought better solutions of the form:

V(frustum) = $h[(a + b)/2]^2 + $ *correction term*   (11.5e)

V(frustum) = $h(a^2 + b^2)/2 + $ *correction term*   (11.5f)

The question is how some OB scribe obtained the correct *correction terms*, and here we can only guess. A possibility that appears to me to be within OB capabilities is the following logical reasoning: The *correction term* must be a volume that goes to zero when $b = a$ because in this case equations (11.5a and 11.5c) are correct. The *correction term* must therefore contain $(a - b)$ as a factor. In order that the term has correct dimensions, the simplest guess is that the term is a constant $\times$ $[h(a - b)^2]$. In order to evaluate the constant, we can require that V(frustum) = V(pyramid) in the limit $b = 0$. In this limit equations (11.5e and 11.5f) become

$$ha^2/4 + kha^2 = ha^2/3 \qquad\qquad (11.5g)$$

$$ha^2/2 + k'ha^2 = ha^2/3 \qquad\qquad (11.5h)$$

Solving for the constants k and k', in these equations yields equations (11.5c and 11.d), the correct solutions for the frustum volume. OB scribes may have empirically found these to be better solutions, but it is doubtful that they could have known them to be exact solutions. Rigorous proofs that these were indeed exact solutions would have to wait more than a millennium for Greek mathematicians.

Thus, correct solutions for frustum volume appear in both OB and XIII dynasty surviving records, but the algorithms, or equivalently their representations as algebraic equations, are different. However, the probable derivations of the OB equations give us a clue to how the Egyptian version, equation (11.4a), was probably derived. Because the term $(a^2 + b^2)$ appears in equation (11.4a), equation (11.4a) can also be looked at as an improvement to the approximation of equation (11.5b). One way of looking at the logic of the derivation is to consider an algorithm change such that

$$V(\text{frustum}) = h(a^2 + b^2)/\ 2 \rightarrow h(a^2 + \text{correction term} + b^2)/3 \qquad (11.6)$$

In order to give the correct answer when V(frustum) = V(prism), *correction term* = a^2 when $a = b$. In order to give the correct answer when V(frustum) = V(pyramid), *correction term* = 0 when $b = 0$. The simplest guess is that *correction term* = ab, which in fact produces equation (11.4a), the *MMP* Problem 14 answer.

Although possible, I think a more probable derivation is based on the geometric visualization of figure 11.11.

The approximate solution of equation (11.5b) can be visualized as approximating the frustum by two prisms. Using the intuitive appreciation that synthesizing a shape with more pieces gives a better approximation, some XIII dynasty mathematician added a third layer. This layer must have an area less that a^2 but more than b^2 and so, perhaps among other choices, he tried ab and found empirically that it was indeed a better approximation. It is doubtful that he could have known that it was the exact solution.

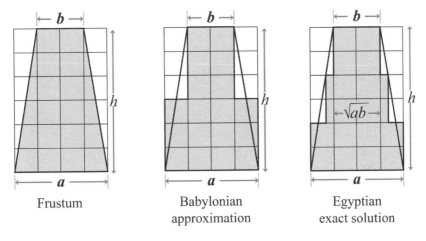

| Frustum | Babylonian
approximation | Egyptian
exact solution |

Figure 11.11 Frustum volume visualizations relevant to *RMP* 14 algorithm

We have now seen how, some four thousand years ago, Egyptian and Babylonian scribes invented plane geometry and solid geometry, and we have seen a representative selection of their contributions. We have also seen something of how the successor Greek civilization treated some of these same problems. We can succinctly contrast the two approaches as *visualization and calculation* vs. *visualization and annotation.* This chapter's consideration of the determination of the volume of a pyramid nicely defines the difference. A scribe in Egypt or Babylon could have said, "To as precisely as anyone knows or cares, the volume of a pyramid is one-third the volume of a prism with the same height and base." Euclid said with certainty, "The volume of a pyramid is one-third the volume of a prism with the same height and base."

Except for the advantage conferred by annotation, was there a change of interest? What was the difference in motivation between the OB scribe who calculated $\sqrt{2}$ to new limits of precision and Archimedes, who calculated π to new limits of precision? Absolutely none. The fact that the Egyptians and the Babylonians only did calculations has given the mistaken impression that their interest was limited to applications. In all civilizations in all eras, mathematicians solve the practical computational requirements of the day, but what motivates them to do mathematics, in all civilizations in all eras, is curiosity, curiosity, curiosity.

Chapter 12

FROM OLD BABYLONIAN SCRIBE TO LATE BABYLONIAN SCRIBE TO PYTHAGORAS TO PLATO

LATE BABYLONIAN (LB) MATHEMATICS

Recovery of documentation of Babylonian mathematics from after the OB era is minimal, but some cuneiform clay tablets have been recovered, mostly from after 500 BCE. Denoted as Late Babylonian (LB), the contents closely resemble the metric geometric algebra of the OB era. Apparently, the intervening millennium was not completely mathematically stagnant because the LB mathematics sometimes exhibits greater sophistication. Despite the hiatus in documentation, there was clearly continuity in pursuit of mathematics because, as previously noted in chapter 7, person-to-person contact was required to communicate nonannotated Babylonian mathematics. Presumably some diminution in mathematics education, combined perhaps with greater destruction in a millennium of constant and more destructive warfare, accounts for the hiatus in mathematical documentation from Mesopotamia.

The LB mathematical revival coincides with and is probably attributable to increased interaction with Greeks. However, whether it was because of infusion of more-curious Greek thought or it was simply a by-product of greater affluence and more commerce, or all of the above, is unanswered. An example of LB mathematics is problem text AO 6484 discussed in appendix B.

Close contact between Greek and Babylonian cultures began after Cyrus of Persia conquered Assyria in 560 BCE and the Persian empire started expanding westward. Egypt, which contained a sizable Greek colony, was conquered by Persia in 525 BCE. By 500 BCE, all of what is present-day Turkey was Persian, which included all of the Ionian Greek areas along the Aegean coast. Greek-Persian relations were sometimes friendly, sometimes hostile, but certainly included cultural interaction. Persian westward expansion ended when Darius was defeated by the Greeks in the battle of Marathon in 490 BCE, possibly the most famous land battle—"for never did so small a band overthrow so numerous a host"—and then when his son Xerxes was defeated at Salamis in 480 BCE, possibly the most famous sea battle.

Alexander the Great's conquest of the Persian empire (333–329 BCE) marked the beginning of Hellenization of the Middle East. After Alexander's death in 323 BCE, his feuding generals split his empire into the Antigonid empire in Greece, the Seleucid empire in Mesopotamia and Persia, and the Ptolemaic Kingdom in Egypt, Libya, and Palestine. With the founding of the Library in Alexandria around 300 BCE, the Ptolemaic Kingdom became an important center for mathematics.

Somewhat strangely, while the LB mathematical revival in what became the Seleucid empire had begun in the sixth century BCE, it was only from the Ptolemaic Kingdom era, more than two hundred years later, that documentation of similar mathematics was recovered from Egypt. Egyptian script, under the influence of Greek alphabetic writing, had changed into a new script referred to as *demotic*. LB mathematics from Egypt is documented on papyri in demotic and Greek. The significance of this apparently delayed development in Egypt is questionable. It may perhaps be nothing more than that the surviving Egyptian record is so lean that the chance discovery of a few documents could completely change the impression that the Egyptian mathematical revival really lagged behind that of Mesopotamia by some two hundred years.

Intriguingly, the Ptolemaic-era Egyptian documentation now includes "quadratic" metric algebra and Pythagorean theorem problems that had not been found in OB-era Egyptian documentation. Whether this reflects continuous mathematical progress in Egypt, parallel to and equivalent to Babylonian progress since OB times, or it just represents recent infusion of LB mathematics from the Seleucid empire has not been resolved yet. Friberg[1] has reviewed demotic and Greek LB mathematics in detail and I shall not consider it further here.

LB mathematics, as OB mathematics, was scribal education, essentially trade-school education. However, as we have amply seen, mathematical curiosity and not just practical calculation are well represented in the surviving OB documentation. Technology changed from Bronze Age to Iron Age during the transition from OB to LB eras, but this placed no significant new demands on scribal mathematics and hence change in scribal mathematics was not great. However, more or less simultaneously with the LB mathematical revival, a more philosophical interest in mathematics began in Greece. This awakening is generally attributed primarily to Pythagoras.

PYTHAGORAS (ca. 580–500 BCE)

The classic Christian concept of creation of our world is given in the King James translation of the book of Genesis:

> *In the beginning God created the heaven and the earth.*
> *And the earth was without form, and void, and darkness was upon the face*
> *of the deep.*

This is creation *ex nihilo*, creation out of nothing, a hard-to-digest concept philosophically, an impossible concept for Albert Einstein. A more recent translation, based on better understanding of the original Hebrew and hence probably more accurate, is that of the Anchor Bible:

> *When God set about to create heaven and earth—the world was a formless*
> *waste.*

The ancient Hebrew creation myth was not *ex nihilo*; creation was pattern out of chaos. It is amusing to realize that a fundamental church dogma is simply a translation error.

The creation myths of ancient Egyptians, Babylonians, and Greeks were also not *ex nihilo*; all were pattern-out-of-chaos myths. They all start with similar concepts, suggesting a common origin, of material entities of the universe having godlike qualities. These entities eventually evolve into the anthropomorphic gods that are more familiar to us, such as Osiris, Marduk, and Zeus. An Egyptian creation myth has imagery reflecting the yearly life-giving flooding of the Nile:

In the beginning there was only water, a chaos of churning, bubbling water called Nu. It was out of Nu that everything began. When the flood receded, out of the chaos of water emerged a hill of dry land, one at first, then more. On this first dry hilltop, on the first day came the first sunrise.

The Genesis story is largely a Babylonian creation myth, *Enuma Elish*, with some monotheistic polish and spin. The correlation is so close that the classic translation and interpretation of the *Enuma Elish* by Heidel[2] is titled *The Babylonian Genesis*. In the *Enuma Elish*, creation starts with intercourse between two oceans:

> *Apsu the primal sweet-water ocean and Timat the salt-water ocean mingled, forming one immense undefined mass that contained all of the elements of which the world would be made.*

A Greek creation myth told by Hesoid also invokes sexual reproduction:

> *In the beginning there was only chaos, abyss.*
> *And Eros, loveliest of all Immortals, who*
> *From the Abyss were born Erebos and dark Night*
> *And Night, pregnant after sweet intercourse*
> *With Erobos, gave birth to Aether and Day*

These are *big bang* myths. They all say that the event most relevant to us is the pattern-out-of-chaos event; there was something before this creation event but not much can be known now about what it was, so largely forget about it and construct a religion based on more anthropomorphic and more appealing gods who evolved after the big bang. Like physicists today, Pythagoras was also attempting to understand the big bang by studying the patterns created out of chaos. Naively but fortunately, he chose to study the patterns formed by numbers and invented number theory.

Today, Pythagoras's fame stems from the mathematics attributed to him. During his lifetime his fame was primarily as the godlike founder of a religious cult, the Pythagorean Brotherhood, whose motto was "All is number." He was not very successful in establishing his cult among the educated populous in his native Samos, so he moved to the Greek colony of Croton in southern Italy where he had more success among the yokels, the sort of tactic favored by evangelical preachers today.

The Pythagoreans apparently lived a monastic life based on moral and ethical behavior. They believed in reincarnation, were vegetarians, and for some reason were forbidden from eating beans. They were also forbidden from wearing wool and wore only simple linen clothing. However, not much is known of their religious rituals and practice because of the cult's strict secrecy rules. Perhaps the motivation for secrecy was to avoid revelation of revolutionary ideas that would be objectionable to religious authority. Lest one think that the Greek religious establishment was receptive to deviant concepts, philosopher/mathematician Anaxagoras was imprisoned (ca. 450 BCE) at Athens for asserting that the sun was not a deity but a huge red-hot stone. Not quite as drastic as the burning at the stake of philosopher/mathematician Bruno in 1600 by the Catholic Inquisition for espousing that Earth circled the sun, but the message was the same. In fact, fear of being labeled heretics was not paranoid; the Pythagoreans at Croton were slaughtered with but a few survivors around 460 BCE.

One of the legends has it that Pythagoras's sojourn in Babylon was as a captive taken during the Persian conquest of Egypt in 525 BCE. Whether or not by a visit to Babylon, voluntary or otherwise, Pythagoras apparently became familiar with LB mathematics. But rather than just continue in the Babylonian style, he went in an entirely new direction: he digitalized LB geometric algebra and converted it into number theory.

"All is number" should really be read as "All is rational numbers," but in ca. 400 BCE Hippasus proved that $\sqrt{2}$ was not a rational number. For the Pythagoreans this was equivalent to a mathematical proof that "Your God is a fake" and legend has it that the immediate response was the conventional response of any religious bureaucracy—they drowned Hippasus. Interestingly, at about this time Pythagoreanism changed to neo-Pythagoreanism with must less mathematical content. The proof that $\sqrt{2}$ is not a rational number plays such an important role in Greek mathematics that I present a detailed, easy-to-understand version of the proof in appendix C.

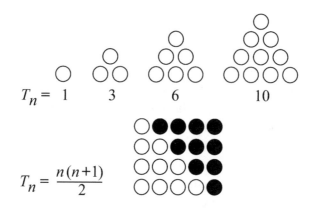

Figure 12.1 Triangular numbers

Figure 12.1 illustrates Pythagoras's "bean counting" visualization of the summation of an arithmetic series,

$$T_n = 1 + 2 + 3 + 4 + \dots n \tag{12.1}$$

Calculating the sum of the arithmetic series is visualized as combining two triangles to form a rectangle of area $n(n + 1)$. We have previously seen in chapter 9 how Egyptian and OB scribes visualized the sum of an arithmetic series in terms of similar geometry; Pythagoras's visualization was also triangular but digitalized. Was Pythagoras's visualization inspired by prior LB mathematics? My guess is, probably, but it is impossible to know.

Figure 12.2 gives the Pythagorean visualization of square numbers. It shows that the sum of the series of odd numbers, visualized as gnomons, is a square number.

$$n^2 = 1 + 3 + 5 + \dots (2n - 1) = \sum_{p=1}^{p=n} (2p - 1) \tag{12.2}$$

The really intriguing aspect of the square number visualization is that Pythagoras is credited with using it to calculate triples.[3] Although it may not at first sight appear to be the same as the Babylonian theorem derivation that was the OB "proof" of the Pythagorean theorem (see chapter 5), it actually is. The three-term equation of Pythagoras (see figure 12.2),

$$(n + 1)^2 = (2n + 1) + n^2 \tag{12.3a}$$

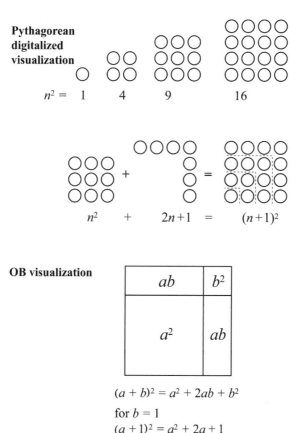

Figure 12.2 Square numbers

will generate triples if $p^2 = (2n + 1)$, which when used in equation (12.3a) yields

$$(p^2 + 1)^2 = 4p^2 + (p^2 - 1)^2 \qquad (12.3b)$$

exactly equation (5.2a) with $Q = 1$, which I conjectured was the way an OB scribe composed triples that could be shown to be right triangles and hence was the OB "proof" of the Pythagorean theorem.

It is interesting to note that it was not until some five hundred years after the death of Pythagoras that the historian Plutarch first attached the name of Pythagoras to the theorem.[4] However, when it was realized that Euclid's proofs were too sophisticated to have been done by Pythagoras at such a primitive stage in Greek mathematics, apparently the simplest proof then known was credited to Pythagoras with only legend as evidence. In fact, he was credited with the proof that could have followed from OB problem text Db$_2$ 146 (see chapter 4). Thus out of thin air is history composed.

Now that much earlier OB use of the theorem is known, there is no need, other than pride and tradition, to give a Greek name to its discoverer. However, crediting Pythagoras with the digitalization considered above appears reasonable because it is not in prior Babylonian records and so it had to be discovered by someone early in Greek mathematical development. The name Pythagoras is the generic name for original, early Greek mathematicians. Since a reasonable "proof" of the Pythagorean theorem follows naturally from consideration of square numbers, it is probable that this was the origin of early Greek awareness of the theorem. Once mathematicians became used to rigorous, Euclidian-type derivations, they lost sight of the possibility that awareness of the Pythagorean theorem did not have to start with a rigorous derivation, and it most probably did not, either in Babylon or in Greece.

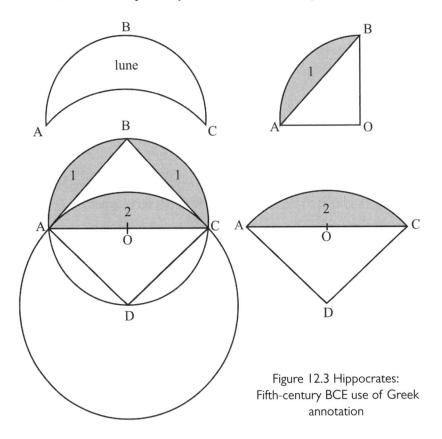

Figure 12.3 Hippocrates: Fifth-century BCE use of Greek annotation

There is good evidence for very early Greek use of the Pythagorean theorem. Hippocrates (ca. 470–410 BCE) (not the Hippocrates of medical fame), who belonged to the Pythagorean cult, used the Pythagorean theorem in his treatment of lunes and segments of circles as illustrated in figure 12.3. This work of Hippocrates is also of particular importance because it is the first documented evidence for addition of alphabetic annotations to geometric drawings. Thus, within about fifty years after the death of Pythagoras, alphabetic annotation was certainly being used and conceivably could even have been invented earlier by Pythagoras. In chapter 7 we saw use of similar notation by Archimedes dated some two hundred years later. There we could see the need for such annotation because of the complexity of the geometry. This problem solved by Hippocrates was too simple to require annotation. I think that the explanation for its introduction lies in the universal and timeless need of scholars to exchange ideas in a Greek world that was widely spread around the Mediterranean. For example, correspondence between Archimedes in Syracuse, Sicily, and Eratosthenes at the Library in Alexandria, Egypt, is documented.

In the Greek era it was difficult to include original drawings in mail, hence the notation had to be not only mathematically unambiguous but it also had to supply instructions on how to draw the required geometric figure. Thus, figure 12.3 possibly never appeared in any original publication. In an era of hand copying of documents there is not much difference between correspondence and publication. In chapter 13 we shall see much more clearly how Euclid's proofs are also drawing instructions. All of the drawings appearing in transcriptions of Euclid, on which all modern translations are based, were possibly not part of the text Euclid published.

FUN QUESTION 12.1: The objective of Hippocrates' construction in figure 12.3 was part of a futile search for a method of squaring the circle because it was not then understood that π was not a rational number. However, his construction did prove that area (segment 2) = twice area (segment 1). Prove it and calculate the area of the lune.

Pythagoras's "All is number" concept may appear to be ridiculous speculation, but it is a seminal attempt at a secular explanation of physical and chemical observations. The world of numbers was observed to produce some remarkable patterns: sums of odd numbers are squares of numbers; multiplying an odd number by an even number converts the odd into an even number; a sum of two squares is another square; $1 + 2 + 3 = 1 \times 2 \times 3 = 6$, a "perfect" number. The real world also is a world of patterns, which was apparently sufficient to convince Pythagoras and his disciples that pursuit of understanding number patterns would lead to understanding the real world.

There were many other attempts by Greeks at secular explanations of physical and chemical phenomena. Perhaps the most familiar is the atomic theory of Democritus (460–370 BCE). Empedocles (495–435 BCE) invented a theory based on the interactions between four basic elements: earth, water, air, and fire. It is Empedocles' theory, embellished with some imaginative mathematics, that became Plato's theory of matter, which we shall next consider as some "comic relief" before returning to the serious work of Euclid in the final chapter.

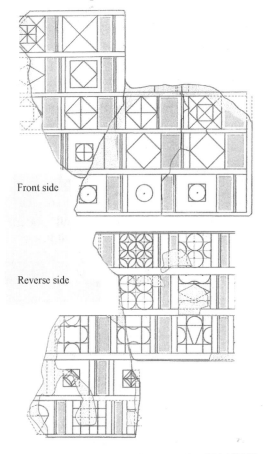

Front side

Reverse side

Figure 12.4 Drawing of OB clay tablet BM 15285

FUN QUESTION 12.2: Figure 12.4, from a recently published drawing and translation of OB clay tablet BM 15285,[5] presents drawing-compass constructions and problems that are hardly less sophisticated than the problem composed by Hippocrates more than a millennium later. Figure FQ 12.2 extracts one of the problems from BM 15285. Calculate the area of the shaded part of the drawing.

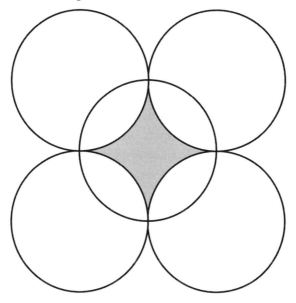

Figure FQ 12.2

Note that this figure has no Greek annotation but was nevertheless solved by some OB scribe, illustrating the point that the annotation was not required to solve such a simple problem. Hence its use by Hippocrates was probably for unambiguous communication.

PLATO (427–347 BCE)

Plato is not known as a great mathematician but as a maker of mathematicians. Inscribed over the doors to his Academy in Athens was the motto "Let no one ignorant of geometry enter here." Plato's Academy surprisingly survived for six hundred years. His original mathematical contributions were apparently minimal, but the consequences of his directing attention to mathematics were great.

Plato's theory of matter, what I call *Platonic chemistry*, is described in his dialog *Timaeus*.[6] Here Plato relates the four basic elements of Empedocles to four of the five three-dimensional figures that can be constructed with equal-sided faces that are illustrated in figure 12.5. Plato did not discover these constructions, but because of his famous use they are frequently referred to as *Platonic solids*. All of these constructions have been credited to Pythagoras, but apparently the octahedron and icosahedron were discovered by Plato's contemporary Theaetetus (417–369 BCE), who also proved that only these five figures with equal-sided faces existed. The fact that there were four basic elements but five figures was close enough for Plato; he made the dodecahedron into the celestial sphere in which the fixed stars were embedded. The four basic elements of Empedocles now had defined geometries amenable to mathematical analysis. Plato combined Pythagoras's geometric number patterns with Empedocles' qualitative conception of real-world chemistry to produce a somewhat quantitative chemical model.

Following Empedocles, Plato believed that these four basic elements combined with each other in varying quantities and configurations to account for the many different observed reactions and materials. However, he divided the faces of the Platonic solids into right triangles as illustrated in figure 12.5. The faces of the dodecahedron he divided into 20 triangles; he did not illustrate his division, but it was probably as illustrated here. Since the dodecahedron has 12 pentagonal faces, $12 \times 20 = 360$, which was apparently a number of celestial significance and there is no further comment about it. He divided the equilateral triangle faces of the tetrahedron, octahedron, and icosahedron into six (1, $\sqrt{3}$, 2) triangles; he divided the faces of the cube into four (1,1, $\sqrt{2}$) triangles. While the Platonic solids by combination and division can make new material, the triangles in the faces cannot change. Let us call the four Platonic solids molecules, and the triangles atoms. Some possible Platonic chemical reactions are

1 water molecule → 5 fire molecules → 2 air molecules + 1 fire molecule,
1 icosahedron   = 5 tetrahedrons   = 2 octahedrons   + 1 tetrahedron,
20 faces × 6 trianges = 5(4 faces × 6 triangles) = 2(8 faces × 6 triangles) + 4
                                                        faces × 6 triangles,
120 atoms   = 120 atom   = 120 atoms

Platonic solids — molecules	Triangles in faces — atoms

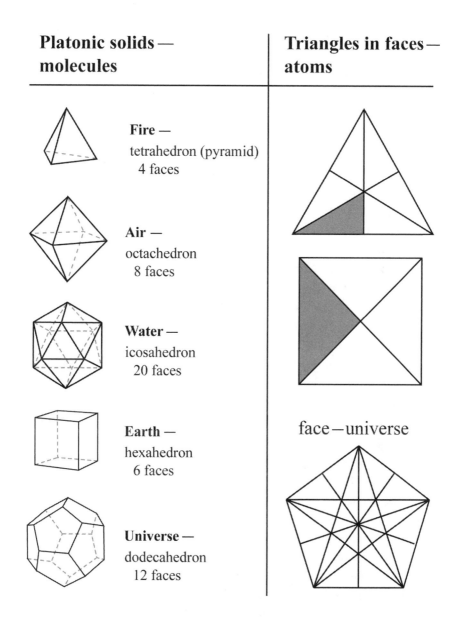

	Fire — tetrahedron (pyramid) 4 faces	
	Air — octachedron 8 faces	
	Water — icosahedron 20 faces	
		face — universe
	Earth — hexahedron 6 faces	
	Universe — dodecahedron 12 faces	

Figure 12.5 Platonic chemistry

The addition of a defined geometry does add a quantitative aspect to the earth, air, water, and fire concept; it even defines reactions in a manner similar to modern chemical notation with a conservation-of-atoms law. Plato considers the earth molecule, the cube, as inert because its triangle does not match and hence cannot react with the other molecules.

Much of Greek philosophy is justifiably criticized as "armchair," implying that it is just speculation unrelated to observation. Plato's chemistry does relate to observed behavior and does indeed account for observed behavior without requiring intervention of gods, although it is based on some convenient assumptions. (In addition to the disposal of the dodecahedron misfit, why is each face divided into six triangles, and not into twelve triangles, or even into an infinite number of similar triangles as we saw was possible in figure 9.2?) However, it fails the basic requirement of a good theory; it has no predictive value, which Plato would have required of any purely mathematical theory.

While predictive value is a necessary condition, it has not always been recognized that it is not a sufficient condition. Recall Ptolemy's flawed theory of planetary motion discussed in chapter 6; because it tracked periodic motion, its ability to predict was outstanding. Plato's setting of the bar lower for theories about the real world than for theories about numbers was actually an insightful departure from the simplistic Pythagorean belief in a one-for-one correspondence between number patterns and the patterns of the real world. The means of observation were just not sufficient at the time to define a good theory and would not be for another two millennia. Plato's theory translates easily into modern molecular and atomic terms, but it is doubtful that it had any influence on modern chemical concepts that were dictated by contemporary experimental observation and that required no clues from Plato.

The requirement that a good theory have predictive value was not too sophisticated a concept for Plato's day. It was already intuitively required to validate religion and none more so than the religion of the Greeks. At the Oracle of Delphi and at other sites predictions were supplied. How could an oracle maintain credibility with predictions that could not possibly have been god-given accurate? Ambiguity, ambiguity, ambiguity, the perennial defense of mystics, astrologers, fortune-tellers, and other charlatans. One of the most famous predictions of the Delphic Oracle was given to Croesus, king of Lydia (in present-day western Turkey) in 560 BCE.[7] In answer to his question about what would be the result if he went to war to eliminate Persian domination, the oracle answered that if he did, he would destroy a great empire. Croesus interpreted this as a positive sign and waged war against King Cyrus. The prediction was validated, but the empire that was destroyed was his.

The "prophet motive" also has a major role in the Hebrew Bible; however, the excuse for inaccurate prediction is generally that too many Jews are not adequately obeying God's commandments and therefore He is no longer required to honor His promise. Christians protect the sanctity of the predictions in the book of Revelations by adding ambiguity of time to ambiguity of meaning as a fail-safe defense.

Only mathematics requires no-excuses prediction. For the given assumptions, proof must be perfect. Whether or not the given assumptions relate to the real world is now understood as a physics, not a mathematics, problem. And thus we come full circle back to Pythagoras's "All is number" theory. He, like Plato, thought he was solving a physics problem. In fact, he referred to himself as a philosopher, a term he coined that literally means a lover of learning, rather than as a mathematician. Since Pythagorean number theory is based on geometric visualization, it is now referred to as Pythagorean geometric algebra. It may have been naive for Pythagoras to have assumed that his geometric discoveries would provide explanations about nature, but it has turned out to have been a prescient conjecture. Now, some 2,500 years later, geometry has indeed become the basis of the fundamental understanding of nature. The geometry of space-time is Einstein's relativity theory, elementary particles are now defined by multidimensional geometries,[8] and the evolution of living organisms and geologic formations are now understood in terms of Mandelbrot's fractal geometry.[9]

Chapter 13

EUCLID

WHO WAS EUCLID?

U p until about the middle of the twentieth century, high school students worldwide knew that Euclid was the Greek mathematician who had written their geometry textbook some two thousand years before. The book was probably titled something like *Euclid's Elements* and it was more or less a literal translation of proofs written by Euclid. This meant that it was written in Greek alphabetic notation, which we first encountered here in chapter 7 in Archimedes' determination of the value of π, and also in chapters 8, 9, 11, and 12. While the invention of such notation had enabled Greek mathematical progress, for most modern high school students it complicated the proofs beyond comprehension, although it presented an interesting challenge to the mathematically gifted. Thus, around the middle of the twentieth century, literal translations of Euclid's proofs were replaced by geometry problems, largely in modern algebraic notation, and the texts had titles such as *plane geometry*. However, considerable use of Greek alphabetic notation has been retained because of its ability to unambiguously define geometric figures. However, the name Euclid has generally disappeared from geometry textbooks with typical bureaucratic throwing-out-the-baby-with-the-bathwater, thus we now have generations who do not recognize the name of the man who wrote a textbook that has been used continuously for some two millennia and except for some change in notation wrote the geometry now studied.

Very little is known about Euclid's life beyond that, apparently after he had gained a reputation as "Euclid the Geometer" at Plato's Academy in

Athens, he was invited to found the school of mathematics and to teach at the Library of Alexandria around 300 BCE. Two legends attest to the fact that he took his mathematics very seriously:[1] When Ptolemy (king of Egypt) asked him if there was in geometry any shorter way than that of *The Elements*, he answered that "there is no royal road to geometry." The second Euclid legend is that when a student who was studying geometry asked him what he would get by learning such things, Euclid said to his slave, "Give him threepence, since he must make gain out of what he learns."

Because Euclid so thoroughly summarized Greek mathematics, copying of earlier work largely ceased. However, no original copies of *The Elements* have survived and translations of Euclid largely trace back to the Greek mathematician Proclus (410–485 CE). Despite the fact that Proclus lived some seven hundred years after Euclid, he had access to a no-longer-extant history of Greek mathematics by Eudemus (270–300 BCE), a contemporary of Euclid. Thus, what little that is known about Euclid and pre-Euclid Greek mathematicians is largely due to Eudemus via Proclus. Transmission of Euclid to Europe was largely by Arabic translations of Proclus that began to be made in the eighth century.

Euclid comes to us today with commentaries spanning some two millennia by literally hundreds of very competent mathematicians. There does not appear to be any detail, no matter how trivial, that has not been exhaustingly critiqued. However, most of these commentaries on Euclid were made prior to the relatively recent appreciation of OB-era mathematics. As we have already seen, even before the invention of Greek alphabetic notation the Pythagoreans had added new depth to OB-era mathematics. However, without the addition of alphabetic notation, Greek mathematics probably would also have quickly stagnated as its antecedent OB-era mathematics had. I shall briefly revisit Euclid in terms of how his work differs from OB geometry beyond its obvious ability to deal with more complex problems that Greek alphabetical notation enabled.

Euclid realized that mathematical proofs must start with some intuitively obvious but nonprovable assumptions. He chose a minimal set of assumptions and required that all his proofs depend only on these assumptions. With these few intuitively obvious assumptions he then rigorously proved some nonobvious geometric results, some important and some trivial. Euclid first stated *propositions* (what was to be proved) and then proved them using only his minimal set of intuitively obvious assumptions. The assumptions plus prior proofs are *the elements* with which Euclid constructed his proofs.

The *axiomatic* (the logical construction of proofs from *the elements*) character of Euclid's proofs was their hallmark. However, Euclid was certainly not the originator of axiomatic proofs. Much earlier, Hippocrates (470–410 BCE) had written *Elements of Geometry*, which Euclid, to some unknown degree, incorporated into his book. In chapter 12 we have previously seen such a striking similarity between OB and Hippocrates' geometric constructions that we can hardly deny OB origins. Can the axiomatic approach to geometry, which has heretofore been considered so characteristically Greek, also be traced back to the OB era?

The surviving OB-era mathematical record contains only numerical calculations from which modern mathematicians have been able to extract the algorithms employed. Since no proofs are explicitly stated, although they may have been orally discussed far beyond the limits of our hearing, we do not think of axiomatic proofs when we consider OB mathematics, but they must underlie all logical mathematical derivations. The question is not whether there were OB axioms; the question is only how OB intuitively obvious assumptions may have differed from Euclid's intuitively obvious assumptions. Euclid's proofs are called rigorous; OB proofs have been called "naive."[2] Certainly OB mathematics was less complex than Euclidian mathematics because of its lack of alphabetic notation, but how was it more naive and how did it otherwise differ from Euclidian geometry? To answer this, let us consider in some detail a few Euclidian proofs.

EUCLID I-I

Let us start with the simplest of Euclid's proofs, how to construct an equilateral triangle that is reproduced in figure 13.1a. There are no prior propositions so all of the elements of this proof are intuitively obvious assumptions. It is easy to read this proof and every line of text conforms to our and Euclid's conception of an intuitively obvious assumption. Note that the drawn figure is superfluous; the construction is adequately described by the notation.

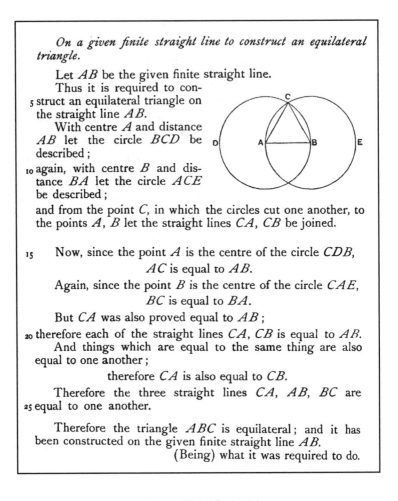

On a given finite straight line to construct an equilateral triangle.

Let AB be the given finite straight line.

Thus it is required to con-
5 struct an equilateral triangle on
the straight line AB.

With centre A and distance
AB let the circle BCD be
described;
10 again, with centre B and distance BA let the circle ACE
be described;

and from the point C, in which the circles cut one another, to
the points A, B let the straight lines CA, CB be joined.

15   Now, since the point A is the centre of the circle CDB,
                    AC is equal to AB.

Again, since the point B is the centre of the circle CAE,
                    BC is equal to BA.

But CA was also proved equal to AB;
20 therefore each of the straight lines CA, CB is equal to AB.

And things which are equal to the same thing are also
equal to one another;
                    therefore CA is also equal to CB.

Therefore the three straight lines CA, AB, BC are
25 equal to one another.

Therefore the triangle ABC is equilateral; and it has
been constructed on the given finite straight line AB.
                            (Being) what it was required to do.

Figure 13.1a Euclid I-1

Although simple, Euclid I-1 is a very practical construction, but it is not an obvious visualization. (I use it to accurately draw equilateral triangles with my computer's PAINT accessory. Would I ever have thought of the method if I had not seen Euclid I-1? Perhaps not.) Apparently the OB construction was as given in figure 13.1b. In a square composed of four squares, half-equilateral triangles, $(1, \sqrt{3}, 2)$ triangles, were constructed in a small square by measuring off a base that is $1/\sqrt{3}$ the side of the small squares. Perhaps this construction is exhibited in OB text problem BM 15285, figure 12.4 (second row, left). Since $\sqrt{3}$ can only be approximately calculated, the construction is only approximate. This is where Euclid's construction differs fundamentally from the OB method. Euclid was aware that $\sqrt{3}$ is an irrational number (see appendix C) and cannot ever be perfectly defined by a measured length. Thus, all of his proofs are limited to figures that can be constructed with only a compass and a straightedge (not a ruler with markings), which assures that approximate measurements are never part of his constructions. OB scribes were certainly not aware of the irrational number concept.

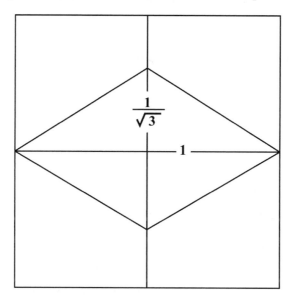

Figure 13.1b OB construction of an equilateral triangle

EUCLID I-22

Euclid I-22 proves that in order for three lengths to form a triangle it is necessary that the sum of the two shorter sides be greater than the longer side. In the composition of FUN QUESTION 4.1, I had assumed that such a condition was intuitively obvious, as I am sure it was to Euclid also. However, Euclid's agenda was to keep the number of intuitively obvious assumptions to a minimum. Since Euclid could prove proposition I-22, he did so and did not count it as an intuitively obvious assumption, even though it was. But when we consider whether OB scribes rigorously proved something, we cannot expect that their agenda also required such austere minimization of assumptions. OB use of intuitively obvious assumptions that are not enumerated by Euclid as such does not constitute lack of rigor.

Euclid's agenda of minimizing the number of intuitive assumptions was sound scientific intuition, consistent with the Occam's razor concept enunciated in chapter 6. Whether this was an original contribution of Euclid or he was using a previous Greek or even OB-era concept is unknown. Minimizing the number of intuitively obvious assumptions creates extra steps in proofs. Thus, not adhering to Euclid's agenda can be a conscious choice that is not necessarily either less rigorous or less sophisticated. For example, Archimedes did not adhere to Euclid's stricture that only constructions with unmarked straight edges are acceptable in his construction that trisected an angle.

Euclid I-22 is presented in figure 13.2. It is fairly simple and easy to understand. (I have not composed any FUN QUESTIONS for this chapter because I think that following a Euclidian proof in Greek notation provides enough fun.) It is interesting to note that in this proposition, Euclid essentially used some modern algebraic notation: the line segment *DF* is also called *A*, *FG* is also called *B*, and *GH* is also called *C*. If he had used lowercase single letters he would have invented modern algebraic notation. In books V–VIII he again used such single-letter notation. I can only guess why there was only such sporadic use. One possibility is that since much of Euclid's propositions are copied from various sources, he simply used the original notations of the various sources. Another possibility is that, as previously noted, no original copies of Euclid's original text survive and the change in notation may be due to some translator's attempt at a proof that was easier to understand.

Out of three straight lines, which are equal to three given straight lines, to construct a triangle : thus it is necessary that two of the straight lines taken together in any manner should be greater than the remaining one.

Let the three given straight lines be A, B, C, and of these let two taken together in any manner be greater than the remaining one,

namely A, B greater than C,

       A, C greater than B,

and B, C greater than A ;

thus it is required to construct a triangle out of straight lines equal to A, B, C.

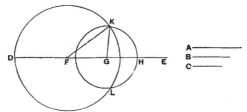

Let there be set out a straight line DE, terminated at D but of infinite length in the direction of E,

and let DF be made equal to A, FG equal to B, and GH equal to C.

With centre F and distance FD let the circle DKL be described ;

again, with centre G and distance GH let the circle KLH be described ;

and let KF, KG be joined ;

I say that the triangle KFG has been constructed out of three straight lines equal to A, B, C.

For, since the point F is the centre of the circle DKL,

     FD is equal to FK.

But FD is equal to A ;

    therefore KF is also equal to A.

Again, since the point G is the centre of the circle LKH,

     GH is equal to GK.

But GH is equal to C ;

    therefore KG is also equal to C.

And FG is also equal to B ;

therefore the three straight lines KF, FG, GK are equal to the three straight lines A, B, C.

Therefore out of the three straight lines KF, FG, GK, which are equal to the three given straight lines A, B, C, the triangle KFG has been constructed.

Figure 13.2 Euclid I-22

EUCLID I-37

Euclid I-37 proves that the area of a triangle with a given base and a given altitude is the same whatever the shape. This is a rigorous proof of a result that was surely already known to Euclid, which can be said for most, if not all, of Euclid's proofs. Figure 13.3a presents Euclid's proof. This proof includes references to both intuitively obvious assumptions and prior propositions. However, to understand the proof it is only necessary to accept that all statements are correct and apply the constructions.

Triangles which are on the same base and in the same parallels are equal to one another.

Let ABC, DBC be triangles on the same base BC and in the same parallels AD, BC;

5 I say that the triangle ABC is equal to the triangle DBC.

Let AD be produced in both directions to E, F;

through B let BE be drawn parallel to CA,

10 and through C let CF be drawn parallel to BD.

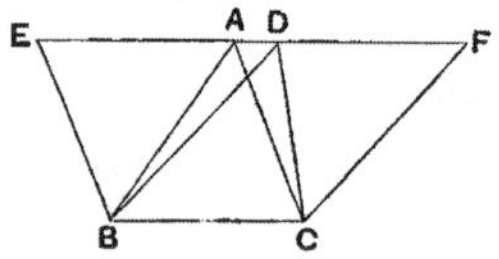

Then each of the figures $EBCA$, $DBCF$ is a parallelogram;

and they are equal,

15   for they are on the same base BC and in the same parallels BC, EF.

Moreover the triangle ABC is half of the parallelogram $EBCA$; for the diameter AB bisects it.

And the triangle DBC is half of the parallelogram $DBCF$;

20 for the diameter DC bisects it.

[But the halves of equal things are equal to one another.]

Therefore the triangle ABC is equal to the triangle DBC.

Therefore etc.

Figure 13.3a Euclid I-37

This is not a very difficult proof, although the visualization used may be a bit unexpected. Conceptually simpler, alternative, equally rigorous proofs are given in figure 13.3b. The upper proof requires defining an equation for the area of a right triangle, which is not required in Euclid's proof. Euclid's proof is *elegant*, much prized by mathematicians, but like *beauty* is hard to define. The lower proof in figure 13.3b uses the fundamental assumption of calculus, that any area can be adequately approximated by an infinite number of infinitesimally small pieces and that simply sliding these pieces relative to each other does not change the area (the deck-of-cards visualization noted in chapter 11) that may have been appreciated by OB-era scribes. It is also an *elegant* proof but was just not included in Euclid's minimized list of obvious assumptions. I have chosen to present this particular proposition because it plays an important role in Euclid's proof of the Pythagorean theorem.

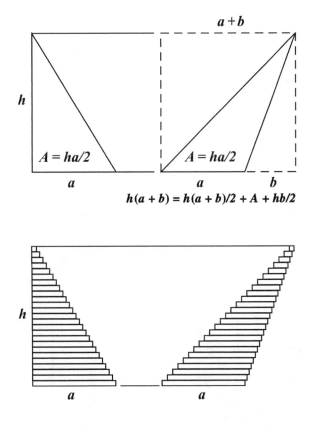

Figure 13.3b Alternative proofs of Euclid I-37

EUCLID I-47: THE PYTHAGOREAN THEOREM

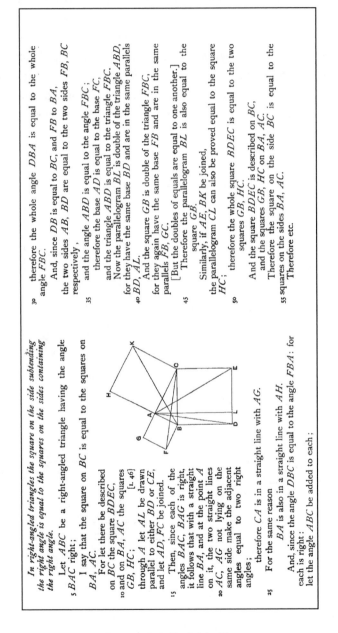

In right-angled triangles the square on the side subtending the right angle is equal to the squares on the sides containing the right angle.

Let ABC be a right-angled triangle having the angle BAC right;

I say that the square on BC is equal to the squares on BA, AC.

For let there be described on BC the square BDEC, and on BA, AC the squares GB, HC; [I. 46] and through A let AL be drawn parallel to either BD or CE, and let AD, FC be joined.

Then, since each of the angles BAC, BAG is right, it follows that with a straight line BA, and at the point A on it, the two straight lines AC, AG not lying on the same side make the adjacent angles equal to two right angles;

therefore CA is in a straight line with AG.

For the same reason BA is also in a straight line with AH.

And, since the angle DBC is equal to the angle FBA: for each is right: let the angle ABC be added to each;

therefore the whole angle DBA is equal to the whole angle FBC.

And, since DB is equal to BC, and FB to BA, the two sides AB, BD are equal to the two sides FB, BC respectively,

and the angle ABD is equal to the angle FBC; therefore the base AD is equal to the base FC, and the triangle ABD is equal to the triangle FBC.

Now the parallelogram BL is double of the triangle ABD, for they have the same base BD and are in the same parallels BD, AL.

And the square GB is double of the triangle FBC, for they again have the same base FB and are in the same parallels FB, GC.

[But the doubles of equals are equal to one another.]

Therefore the parallelogram BL is also equal to the square GB.

Similarly, if AE, BK be joined, the parallelogram CL can also be proved equal to the square HC;

therefore the whole square BDEC is equal to the two squares GB, HC.

And the square BDEC is described on BC, and the squares GB, HC on BA, AC.

Therefore the square on the side BC is equal to the squares on the sides BA, AC.

Therefore etc.

Figure 13.4a Euclid I-47

Figure 13.4a presents one of the two proofs that Euclid gives of the Pythagorean theorem. It is his preferred proof because he thought it required simpler, more obvious, intuitive assumptions. At first sight, understanding this proof is a daunting challenge. When it was taught as a rote exercise, as it too frequently was by lazy or incompetent teachers, it was a painful and useless lesson. Euclid did not perceive this proof, or any other proof, in the line-by-line sequence in which it finally appears. Rather, there was an overall visualization and then he went back and filled in the details. Once Euclid's visualization is understood, this also becomes simple to understand. Euclid's overall visualization must have been more or less as presented in figure 13.4b.

This proof is again of a previously known but possibly not rigorously proved result, $a^2 + b^2 = c^2$, defined by the first image.

- In the second image a line is drawn from point A that divides the c^2 square into two rectangles. It appears likely that the left rectangle has an area equal to a^2 and the right rectangle has an area equal to b^2, and this is what Euclid seeks to prove.
- In the third image the a^2 square is divided into two equal triangles.
- In the fourth image a corner of the triangle (shaded) is dragged from A to C, parallel to its base FB. By Euclid I-37 this does not change its area.

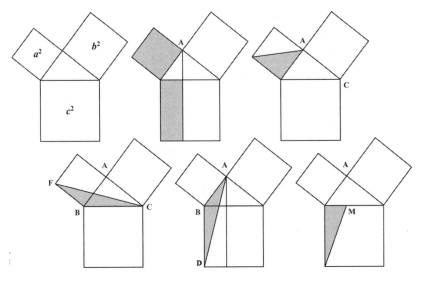

Figure 13.4b Visualization of Euclid I-47

- In the fifth image the triangle is right-rotated 90°.
- In the sixth image a corner of the triangle is dragged from A to M parallel to its base BD. By Euclid I-37 this does not change its area.

Since half of a^2 exactly fills half of the left rectangle, a^2 exactly fills the left rectangle and similarly it is proved that b^2 exactly fills the right rectangle. QED.

This rigorous proof of "For every right triangle of sides (a, b, c), $a^2 + b^2 = c^2$," is the next-to-last proposition in Euclid's book I. The last proposition, Euclid I-48, is its inverse: "Every triangle of sides (a, b, c), for which $a^2 + b^2 = c^2$ is a right triangle."

In previous chapters we have seen various instances where OB solutions of problem texts could be interpreted as Euclid-like rigorous proofs. It has been impossible to conclude with any certainty that OB scribes appreciated existence of such proof aspects of their numerical solutions, but even if they did the proofs were simply accidental encounters of the mathematical kind. In contrast, Euclid's proofs represent a systematic attempt at understanding the rules that govern geometric figures.

EUCLID II-22

The previously considered propositions are rigorous proofs of already known and useful constructions. Euclid II-22 rigorously proves that when a four-sided rectilinear figure is inscribed in a circle, the sum of opposing angles is equal to 180° as shown in figure 13.5. This result was probably not known prior to Euclid's proof, or prior to proof by whomever Euclid copied it from, although looking at figure 13.5, an obtuse angle is always paired with an acute angle, so the result might have been guessed. Possibly there has never been any practical use for this result. So why did Euclid prove it? My guess is because he could, for the sheer joy of rigorously proving a guess. This and much of Euclid has a recreational math aspect and is why many people enjoy working through the difficult language and the difficult notation of Euclid's proofs. Much of mathematics has a recreational aspect. It is an infinitesimally fine line that separates serious mathematics from recreational mathematics.

I shall not explain Euclid's proof here. It requires too many elements that I have not proved previously, but the result is shown in figure 13.5 in algebraic notation. It is an elegant proof, and using algebraic notation is not difficult. It is a shame how modern mathematics teachers have made what should and could have been a fun subject into an incomprehensible one.

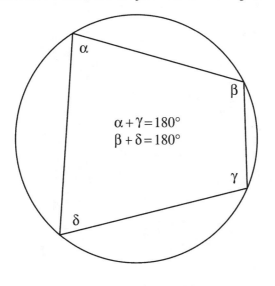

Figure 13.5 Euclid II-22

EUCLID II-14: THE BABYLONIAN THEOREM

This book started with the Babylonian theorem discovered by an OB scribe and it will appropriately end with Euclid's treatment of the same problem some 1,500 years later. Euclid's proof is given in figure 13.6a. The first eleven lines of the proof are just to show that the rectangle BEDC can represent the area of any rectilinear figure; the remaining steps prove that the area of the rectangle is equal to the area of a square with sides equal to *HE*, so this is really a squaring-the-rectangle construction. In fact, it is exactly the same as the eighth-century BCE Hindu squaring-the-rectangle construction of figure 5.5.

Euclid II-14 is not an easy proof to follow in Greek notation and I will not attempt to interpret Euclid's proof in detail, except to note that line 23 defines use of the Pythagorean theorem, Euclid I-47. To appreciate how Euclid's construction relates to the Babylonian theorem, the construction is presented again in algebraic notation in figure 13.6b.

To construct a square equal to a given rectilineal figure.

Let A be the given rectilineal figure;
thus it is required to construct a square equal to the rectilineal figure A.

5 For let there be constructed the rectangular parallelogram BD equal to the rectilineal figure A.

Then, if BE is equal to ED, that which was enjoined will have been done; for a square BD has been constructed equal to the rectilineal figure A.

10 But, if not, one of the straight lines BE, ED is greater.
Let BE be greater, and let it be produced to F;
let EF be made equal to ED, and let BF be bisected at G.
With centre G and distance one of the straight lines GB, GF let the semicircle BHF be described; let DE be produced
15 to H, and let GH be joined.

Then, since the straight line BF has been cut into equal segments at G, and into unequal segments at E,
the rectangle contained by BE, EF together with the square on EG is equal to the square on GF.

20 But GF is equal to GH;
therefore the rectangle BE, EF together with the square on GE is equal to the square on GH.
But the squares on HE, EG are equal to the square on GH;
25 therefore the rectangle BE, EF together with the square on GE is equal to the squares on HE, EG.
Let the square on GE be subtracted from each;
therefore the rectangle contained by BE, EF which remains is equal to the square on EH.
30 But the rectangle BE, EF is BD, for EF is equal to ED;
therefore the parallelogram BD is equal to the square on HE.
And BD is equal to the rectilineal figure A.
Therefore the rectilineal figure A is also equal to the square
35 which can be described on EH.
Therefore a square, namely that which can be described on EH, has been constructed equal to the given rectilineal figure A.

Figure 13.6a Euclid II-14

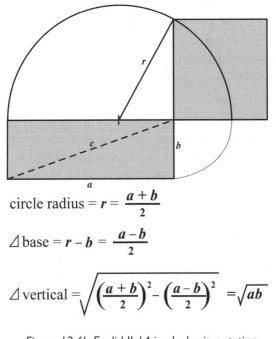

$$\text{circle radius} = r = \frac{a+b}{2}$$

$$\triangle \text{ base} = r - b = \frac{a-b}{2}$$

$$\triangle \text{ vertical} = \sqrt{\left(\frac{a+b}{2}\right)^2 - \left(\frac{a-b}{2}\right)^2} = \sqrt{ab}$$

Figure 13.6b Euclid II-14 in algebraic notation

Now we can see that Euclid had constructed the Babylonian theorem:

For any right triangle (*a*, *b*, *c*) it is possible to construct another right triangle with sides $\sqrt{4ab}$, $a - b$, and $a + b$.

Euclid's proof of the Babylonian theorem depended on prior proof of the Pythagorean theorem, while the OB proof of the Pythagorean theorem depended on prior proof of the Babylonian theorem. To understand this apparent paradox, let us briefly reprise the OB process: An accidental observation of the pattern of figure 4.2 inspired composition of an algorithm that we now define in algebraic notation by $4ab = (a + b)^2 - (a - b)^2$. It was intuitively obvious that this equation exactly defined figure 4.2. This would have been just as intuitively obvious to Euclid, but he would not have included it among his intuitively obvious assumptions simply because he could have proved it with his smaller set of assumptions. But there was nothing naive about the OB derivation of this equation.

Integer solutions to this equation were observed to produce Pythagorean triples (right triangles with integer sides) and thus was the Babylonian theorem "proved." This result was then generalized to include nonintegral solutions also and thus was the Pythagorean theorem "proved."

The OB "proofs" were inductive. Because the set of numbers is infinite, no matter how many numerical fits one can observe it must be just a negligible fraction of the total number of possible fits. Induction produces conjectures; deduction produces theorems. We can now legitimately refer to Euclid II-14 as proof of the Babylonian theorem. Similarly, we can now legitimately refer to Euclid I-47 as proof of the Pythagorean theorem, but until at least after the death of Pythagoras, ca. 500 BCE, it was probably only the Pythagorean conjecture. There is nothing naive about inductive reasoning. It can be useful and it is still used by mathematicians until a deductive proof can be derived. It is the starting point for deductive proofs.

I have previously noted two possible rigorous OB proofs of the Pythagorean theorem, in the solution of problems texts Db_2 146 in chapter 4 and IM 5547 in chapter 8. But what is possible does not count and until some new evidence is found, what can legitimately be called the Babylonian theorem and the Pythagorean theorem after Euclid II-14 and I-47 were probably just conjectures in the OB era.

We have now seen a representative sampling of OB-era metric geometric algebra and Greek nonmetric geometric algebra. How do they compare? Greek progress in geometry did not end with Euclid. Apollonius (262–190 BCE) added analysis of conic sections, but if we exclude this and other later contributions, OB-era scribes and Euclid related to similar geometry. Since any rectilinear figure can be subdivided into just triangles, the plane geometry visualized by OB-era scribes and by Euclid was essentially the same, figures defined by triangles and circles. However, the only triangles calculable by OB-era scribes were right triangles and they usually further limited consideration only to Pythagorean triples because that made calculation easier. Euclidian geometry was able to consider triangles of any shape and hence Euclidian geometry also deals with relationships between angles, a consideration that is largely absent in OB-era geometry.

OB mathematics was basically scribal-school training for practical applications; universal, timeless mathematical curiosity extended the range and depth. Greek mathematics began with Pythagoras's most ambitious conceivable motivation, belief that his mathematics would enable understanding of the creation of the universe (or that was at least the marketing ploy for his religious cult). Plato reduced this ambition to the slightly more reasonable objective that it would enable understanding chemistry, but Euclid was apparently satisfied with simply discovering the rules of geometry.

The ability of nonmetric geometry to consider properties of figures without requiring numerical calculations allowed Euclidian geometry to obtain broader and deeper understanding. But taking this limitation into account, OB-era scribes were no less curious, inventive, or logical. But the invention of nonmetric alphabetic notation also enabled exchange of ideas between mathematicians scattered throughout the Greek-speaking world, the cross-fertilization of which surely played a huge part in Greek mathematical progress, just as it plays a huge part in scientific progress today. In fact, the need to exchange ideas probably inspired the invention of Greek alphabetic notation, just as physicists' need for rapid, global exchange of ideas and data inspired the invention of the Internet.

APPENDIX A
ANSWERS TO
FUN QUESTIONS

1.1: Write the base-5 number 234 as a base-10 number.

$234_5 = 2 \times 5^2 + 3 \times 5 + 4 = 69_{10}$

1.2: Write the base-10 number 189 as a base-5 number.

Hint: $189 = 125a_3 + 25a_2 + 5a_1 + a_0$. Start by finding the largest possible value for a_3. This algorithm is called a *greedy algorithm*. We naturally and intuitively use this algorithm every time we do a pencil/paper long division.

$a_3 = 1 \Rightarrow 189 - 125 = 64 = 25a_2 + 5a_1 + a_0 \Rightarrow a_2 = 2 \Rightarrow 64 - 50 = 16$
$= 5a_1 + a_0 \Rightarrow a_1 = 3 \Rightarrow 16 - 15 = 1 = a_0$
$189_{10} = 1{,}231_5$

1.3: Write the base-5 fractional number 0.234 as a base-10 number.

$0.234_5 = 2 \times 5^{-1} + 3 \times 5^{-2} + 4 \times 5^{-3} = (2 \times 25 + 3 \times 5 + 4)/125 =$
$69/125 = 0.552_{10}$

1.4: Write the base-10 fractional number 0.189 as a base-5 number.

$0.189_{10} = a_{-1}5^{-1} + a_{-2}5^{-2} + a_{-3}5^{-3} + a_{-4}5^{-4} \Rightarrow a_{-1} = 0 \Rightarrow 0.189 = a_{-2}5^{-2} +$
$a_{-3}5^{-3} + a_{-4}5^{-4} \Rightarrow a_{-2} = 4 \Rightarrow 0.189 - 0.16 = 0.029 = a_{-3}5^{-3} + a_{-4}5^{-4} \Rightarrow a_{-3} = 3 \Rightarrow 0.029 - 0.024 = 0.005 = a_{-4}5^{-4} \Rightarrow a_{-4} = 30.189_{10} \cong 0.0433_5$

2.1: Using Egyptian multiplication, calculate 83×97. Compare the number of operations required for this calculation with the number of operations required to do a modern pencil/paper calculation with a memorized multiplication table.

step	operation	\97
1	97 + 97 = 194	\2×97
2	194 + 194 = 388	4×97
3	388 + 388 = 776	8×97
4	776 + 776 = 1,552	\16×97
5	1,552 + 1,552 = 3,104	32×97
6	3,104 + 3,104 = 6,208	\64×97
7,9	6,208 + 1,552 + 194 + 97 = 8,051	83×97

Modern pencil/paper multiplication requires seven operations, four multiplications and three additions:

$$
\begin{array}{r}
97 \\
\times 83 \\
\hline
291 \\
776 \\
\hline
8,051
\end{array}
$$

2.2: Using Egyptian multiplication, calculate 8,051/97. Hint: If multiplication is successive additions, then division is successive subtractions.

The algorithm is exactly the same as the previous multiplication algorithm. Doubling of 97 continues until a sum of doublings equals 8,051, which of course yields 84.

2.3: Write the common fraction 1/11 as a sum of unit fractions with even denominators.

The greediest even unit fraction less than 1/11 is 1/12; 1/11 = 1/12 + 1/132.

3.1: When multiplying using equation (3.1b), it is possible to reduce the required number of squares from 59 to only 30. How?

Using equation (3.1b), calculate $ab/2$ then double this answer by simple addition, which illustrates that less memorization can be obtained by more calculation.

3.2: There is a subtle problem with using equation (3.1b).

- If $(a + b)$ is odd, then $(a - b)$ must also be odd. Prove this.

  For $(a+b)$ to be odd, let a be odd and b be even: $a = (2m + 1)$, $b = 2n$, where m and n are integers.

  Thus $(a + b) = 2(m + n) + 1$, an odd number, and $(a - b) = 2(m - n) + 1$, also an odd number. QED.

- If $(a + b)$ and $(a - b)$ are odd, then $(a + b)/2$ and $(a - b)/2$ are not integers and hence cannot be found in tables of squares of integers up to 59^2. Invent a method that enables use of equation (3.1b) with such tables of squares of integers even when $(a + b)$ is odd.

  Use equation (3.1b) to calculate $a(b - 1)$, which is now an even number. Calculate $ab = a(b - 1) + a$.

3.3: Prove that if R_1 and R_2 are regular sexagesimal numbers, then $R = R_1 R_2$ is also a regular sexagesimal number.

Let $R_1 = 2^i 3^j 5^k$ and $R_2 = 2^l 3^m 5^n$, where i,j,k,l,m,n are integers, then $R = R_1 R_2 = 2^{(i+l)} 3^{(j+m)} 5^{(k+n)}$. QED.

3.4: Is the number 405 a base-10 regular number? Is it a base-60 regular number?

The prime factors of base-10 are 2 and 5. $405/2 \neq$ integer so 405 is not a base-10 regular number.

The prime factors of base-60 are 2, 3, and 5, so 405 is also not a base-60 regular number.

3.5: How would a lazy OB scribe have divided by 49?

He would have obtained an approximate answer by dividing by 48, a close regular number.

4.1: Not only do we not yet know that $(a + b)/2$, $(a - b)/2$, and $\sqrt{ab}$ define a right triangle, we do not even know that they define any triangle. To form a triangle, the sum of the two shortest lengths must be greater than the longest length. Algebraically prove that this is true for these three lengths and hence that they do at least define a triangle.

To prove that $(a + b)/2 < (a - b)/2 + \sqrt{ab} \Rightarrow$ prove that $b < \sqrt{ab}$. Since $a > b$, $b < \sqrt{ab}$. QED.

4.2: OB problem text TMS 1 asks: What is the radius of a circle circumscribed about an isosceles triangle whose altitude is 4 units and whose base is 6 units?

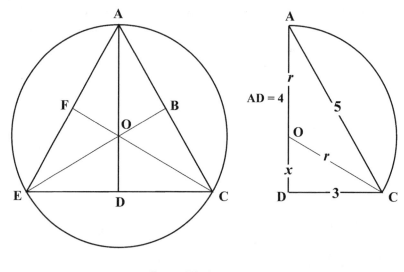

Figure FQ Answer 4.2

To place the center of the circle (point O) equidistant from each corner of the triangle, from the midpoints of each side (points B, D, F) draw perpendiculars that meet at point O. For proof of this construction see Euclid IV-5. The triangle can be divided into two (3, 4, 5) triangles. The simplest solution is to note that triangles ABO and ADC are similar, therefore $r/5 = (5/2)/4$, yielding $r = 25/8$. However, this solution requires use of Euclid IV-5, which was apparently beyond OB capabilities. The less general OB solution, which took advantage of the left-to-right symmetry, intuitively reasoned that the center must lie on the line AD at a distance r from A. Applying the Pythagorean theorem to triangle OCD: $r^2 = x^2 + 3^2$ and using $r + x = 4 \Rightarrow r^2 = (4 - r)^2 + 3^2 \Rightarrow r = 25/8$.

4.3: The Greek mathematician Heron (ca. 100 CE) asked: What is the largest square that can be inscribed in an isosceles triangle whose altitude is 4 units and whose base is 6 units?

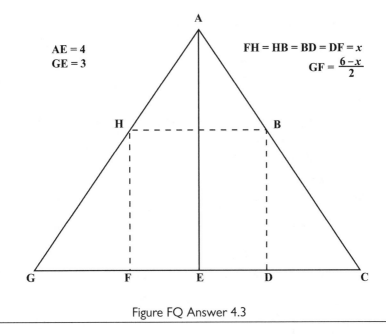

AE = 4
GE = 3

FH = HB = BD = DF = x
$$GF = \frac{6-x}{2}$$

Figure FQ Answer 4.3

Triangles AEG and HFG are similar so that: $2x/(6 - x) = 4/3 \Rightarrow x = 12/5$.

4.4: The sum of the ages of brothers Pat and Mike is 19. The product of their ages is 60. How old are Pat and Mike? Solve this using the method you learned in high school, not by the Babylonian method.

Let Pat be the oldest.

$p + m = 19$, $pm = 60 \Rightarrow p + 60/p = 19 \Rightarrow p^2 - 19p + 60 = 0$. Factoring: $(p - 4)(p - 15) = 0 \Rightarrow p = 15$, $m = 4$.

4.5: Solve equations (4.4) using the figure 4.2 visualization.

$$ab' = 210 \qquad\qquad (4.4a)$$

$a + b' = 29$ (4.4b)

$(a - b')^2 = (a + b')^2 - 4ab' \Rightarrow (a - b') = \sqrt{29^3 - 840} = 1 \Rightarrow a = 15, b' = 14$

4.6: Solve Db$_2$ 146 using only the visualization of figure 4.6.

The area of a rectangle (ab/2) is 0.75 and its diagonal (c) is 1.25. What is its length and width?

Shaded part of figure 4.6: $1.25^2 = (a - b)^2 + 1.5 \Rightarrow (a - b)^2 = 0.0625 \Rightarrow (a - b) = 0.25$

Right triangle $(a, b, c) \Rightarrow$ right triangle $[(a + b), (a - b), 4ab] \Rightarrow (a + b)^2 = (a - b)^2 + 3 \Rightarrow$

$(a+b) = \sqrt{0.0625 + 3} = 1.75$

$[(a + b) + (a - b)]/2 = a = [1.75 + 0.25]/2 = 1$

$[(a + b) - (a - b)]/2 = b = [1.75 - 0.25]/2 = 0.75.$ QED.

4.7: Solve equations (4.7) algebraically.

$-(y_2 - y_1)x + 30y_1 = 840$ (slightly rearranged)   (4.7a)

$y_2 - y_1 = 20$ (4.7b)

$y_2/y_1 = x/(30 - x)$ (4.7c)

$30y_1 - 20x = 840$, using 4.7b in 4.7a to eliminate $(y_2 - y_1)$ (A1)

$y_1 = (600 - 20x)/(2x - 30)$, using 4.7b in 4.7c to eliminate y_2 (A2)

$x^2 + 48x = 1,080$, using A2 in A1 to eliminate y_1

$x^2 + 48x + 21^2 = (x + 21)^2 = 1,080 + 21^2 = 1,521 = 39^2$, "completing the square"

$\Rightarrow x + 21 = 39 \Rightarrow x = 18 \Rightarrow y_1 = 40$ from (A2), $y_2 = 60$ from (4.7b). QED.

4.8: Referring to figure 4.12, show that $x = d/\sqrt{2}$ divides the triangle into two equal areas.

$2A_2 = xy_2 = (A_1 + A_2) = (y_1 + y_2)d/2$

$y_2/x = (y_1 + y_2)d$, similar triangles $\Rightarrow x = d/\sqrt{2}.$ QED.

5.1: Do you remember any elementary physics? As the weight in figure 4.2 is increased, eventually one of the ropes will break. Assuming that all ropes are of equal strength, which rope will break first?

> The tension in the upper pair of ropes is $(5/3)(W/2)$; the tension in the bottom pair is $(5/4)(W/2)$. An upper rope will break first.

5.2: Using equation (5.2a), derive two triples for $Q \neq 1$.

> $P, Q = 9, 4 \Rightarrow P + Q = 12, P - Q = 5, 2\sqrt{PQ} = 13$
>
> $P, Q = 12, 3 \Rightarrow P + Q = 15, P - Q = 9, 2\sqrt{PQ} = 12$

5.3: In the transcription of Plimpton 322 of figure 5.7, except for the (45, 60, 75) triple, when you normalize by dividing by $b = 2\,pq$, a regular number, why do you get a nonterminating fraction. Why do you get a terminating fraction when you normalize the (45, 60, 75) triple?

> In Plimpton 322, $b = 2\,pq$ is a base-60 regular number, but in figure 5.7 the numbers are recorded in base-10. However, only 5 is a factor for (45, 60, 75) and 5 is a prime factor of both 10 and 60.

5.4: In the transcription of Plimpton 322 of figure 5.7, if you normalize any of the huge triangles, you obtain small triangles. Why does that not convert Plimpton 322 into a practical trigonometric table?

> It is not a question of the magnitude of the number but of the number of digits. Thus the number 18,514 in figure 5.7 was too large to measure precisely, but the number 1.8514 requires rulers with finer subdivision than was practical.

5.5: Derive equation (5.3) using the cut-and-paste visualization of figure 4.8.

> That it can be derived by this visualization does not mean that it was. Robson (see Robson reference to chapter 5) has proposed such a derivation. Even if equation (5.3) could have been geometrically derived, there still remain too many sophisticated algebraic steps for this to have been the OB derivation of the pq method.

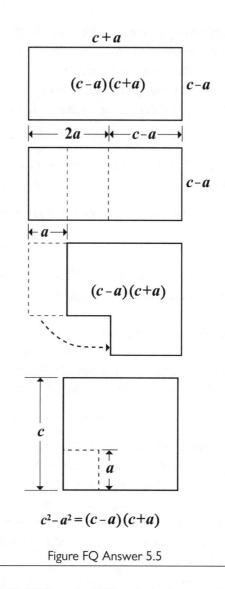

$$c^2 - a^2 = (c - a)(c + a)$$

Figure FQ Answer 5.5

6.1: A problem in VAT 6598 reads, "What is the length of the diagonal of a door of height 40 and width 10?" A student scribe some four thousand years ago could solve this. Can you? Use the cut-and-paste visualization.

The cut-and-paste visualization is equivalent to using equation (6.5a) as:

$c = \sqrt{a^2 + B} \cong a + \dfrac{B}{2a}$, with $B = b^2$, $a = 40$, $b = 10 \Rightarrow c = 40 + 100/80$

$= 41.25$. The exact, electronic calculator solution is $c = 41.23$.

6.2: A door has a height of 5 and a diagonal of 22. What is its width? Use the cut-and-paste visualization.

The cut-and-paste visualization is equivalent to using equation (6.5b) as:

$a = \sqrt{c^2 + B} \cong c - \dfrac{B}{2c}$, with $B = b^2$, $c = 22$, $b = 5 \Rightarrow a = 22 - 25/44 = $

21.43. The exact calculator solution is $c = 21.42$.

6.3 Prove equation (6.10d).

Just simplify equation (6.10d).

7.1: Unlike the scribe from Susa, Archimedes did not obtain a value for π by calculating the area of an n-gon. Can you?

For the inscribed $2n$-gon, the base of each on the $2n$ triangles is given by equation (7.7a),

$s_{2n} = \sqrt{2 - \sqrt{4 - s_n^2}}$

The altitude of each triangle is $h_{2n} = \sqrt{1 - \left(\dfrac{s_{2n}}{2}\right)^2}$. Thus the area of the $2n$-gon is $A_{2n} = 2n(s_{2n}h_{2n}/2) = \pi$. Since $h_{2n} < 1$, the calculated π will be slightly less than Archimedes' calculation, but since $h_{2n \to 1}$ as n increases, both results agree closely for large n.

For the polygon circumscribed about a circle, since the altitude remains constant at 1, the area calculation exactly matches Archimedes' polygon circumference calculation.

8.1: Use the Babylonian result, *for every right triangle the sum of the internal angles is two right angles*, to prove that *the sum of the internal angles of any triangle is equal to two right triangles*, which is just Euclid I-32. We will probably never know whether anyone in OB ever came to this conclusion, but someone certainly could have.

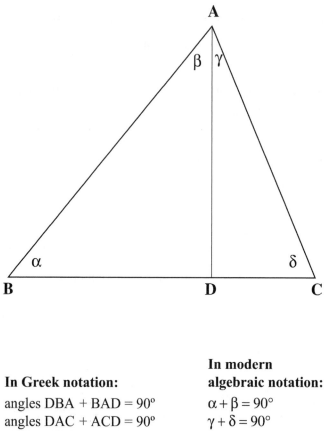

	In modern
In Greek notation:	**algebraic notation:**
angles DBA + BAD = 90°	$\alpha + \beta = 90°$
angles DAC + ACD = 90°	$\gamma + \delta = 90°$
angle BAC = BAD + DAC	
DBA + BAC + ACD = 180°	$\alpha + (\beta+\gamma) + \delta = 180°$

Figure FQ Answer 8.1

9.1: Let the triangle that is divided by an infinite sequence of right triangles, as in figure 9.2, be a 30°, 60°, 90° triangle (half of an equilateral triangle). Let the hypotenuse of triangle ABC be 2. What is the area of A_4?

Areas of similar triangles are related as the squares of the same side.

$$\left(\frac{\sqrt{3}}{2}\right)^2 A_1/ABC = (1/2)^2 = 1/4, \; A_2/A_1 = A_3/A_2 = A_4/A_3 =$$

$$\left(\frac{\sqrt{3}}{2}\right)^2 = 3/4 \Rightarrow A_4/ABC = (1/4)(3/4)^3 = 27/256$$

9.2: Prove that the bases a_i's in figure 9.2 also form a geometric series. For the triangle defined in the previous FUN QUESTION, what is a_{10}?

$a_2/a_1 = a_3/a_2 = \ldots = a_n/a_{n-1} = \left(\dfrac{\sqrt{3}}{2}\right)$, definition of a geometric series.

For $a_1 = 1/2$, $a_n = (1/2)\left(\dfrac{\sqrt{3}}{2}\right)^{n-1}$

$\Rightarrow a_{10} = \dfrac{(\sqrt{3})^0}{2^{10}} = 140.3/1{,}024 = 0.137$

10.1: In figure 10.1 a hexagon is circumscribed about a circle; an equilateral triangle is inscribed in the circle; a square is inscribed in the triangle; a rectangle is inscribed in the square. The rectangle (shaded) has an area $ab = 8$ and $a/b = 4$. What is the area of the trapezoid (shaded)? A good OB student could have solved this some four thousand years ago. Can you?

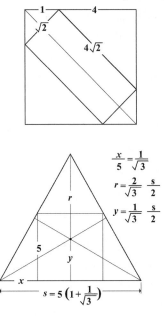

$\dfrac{x}{5} = \dfrac{1}{\sqrt{3}}$

$r = \dfrac{2}{\sqrt{3}} \cdot \dfrac{s}{2}$

$y = \dfrac{1}{\sqrt{3}} \cdot \dfrac{s}{2}$

$s = 5\left(1 + \dfrac{1}{\sqrt{3}}\right)$

Area triangle $= s(r+y)/2$

Area hexagon $= 6r^2/\sqrt{3}$

Area trapezoid $= [6r^2/\sqrt{3} - s(r+y)/2]/3 = 27.9$

Figure FQ Answer 10.1

11.1: Apply the greedy algorithm (see chapter 2) to calculate N(pyramid)/ N(prism) = 30/64 and N(pyramid)/N(prism) = 204/512 in terms of Egyptian fractions.

$64/30 = 2 +$ remainder $\Rightarrow 1/3 =$ greedy subtraction:
$30/64 - 1/3 = (90 - 64)/(3 \times 64) = 26/(3 \times 64) = 13/96$
$96/13 = 7 = 1 +$ remainder $\Rightarrow 1/8 =$ greedy subtraction:
$13/96 - 1/8 = (104 - 96)/(8 \times 96) = 1/96$
N(pyramid)/N(prism) $= 30/64 = 1/3 + 1/8 + 1/96$

$512/204 = 2 +$ remainder $\Rightarrow 1/3 =$ greedy subtraction:
$204/512 - 1/3 = (612 - 512)/(3 \times 512) = 50/(3 \times 256)$.
$(3 \times 256)/50 = 15 +$ remainder $\Rightarrow 1/16 =$ greedy subtraction:
$[(50 \times 16) - (3 \times 256)]/(16 \times 3 \times 256) = 1/348$
N(pyramid)/N(prism) $= 204/512 = 1/3 + 1/16 + 1/348$

11.2: Show that volume equation (11.4b), derived by calculus, is equivalent to the algebraic generalization of the *MMP* 14 calculation of equation (11.4a). The factorization $(a^3 - b^3) = (a - b)(a^2 + ab + b^2)$ converts equation (11.4b) into equation (11.4a).

11.3: Show that equation (11.4e) can be algebraically converted into equation (11.4a).
$(a - b)^2 + 3b(a - b) + 3b^2 = a^2 - 2ab + b^2 + 3ab - 3b^2 + 3b^2 = a^2 + ab + b^2$. QED.

11.4: Show algebraically that equations (11.5c and 11.5d) can be transformed into the *MMP* 14 solution, equation (11.4b).
$[(a + b)/2]^2 + (a - b)^2/12 = [3(a^2 + 2ab + b^2) + (a^2 - 2ab + b^2)]/12 = (a^2 + ab + b^2)/3$. QED.
$(a^2 + b^2)/2 - (a - b)^2/6 = [3a^2 + 3b^2 - a^2 + 2ab - b^2]/6 = (a^2 + ab + b^2)/3$. QED.

12.1: The objective of Hippocrates' construction in figure 12.3 was part of a futile search for a method of squaring a circle (the graphical construction of a square with an area equal to that of a given circle) because it was not then understood that π was not a rational number. However, his construction did prove that area (segment 2) = twice area (segment 1). Prove it and calculate the area of the lune.

Referring to figure 12.3: r = radius of the small circle $\Rightarrow \sqrt{2}\, r$ = radius of large circle

Area of small circle = πr^2, area of square = $2r^2 \Rightarrow$ area seg(1) = $\dfrac{(\pi - 2)}{4} r^2$

Area of sector ACD of large circle = $\pi (\sqrt{2}r)^2/4$, area of triangle ACD = $r^2 \Rightarrow$

Area of seg(2) = (area sector – area triangle) = $\dfrac{(\pi - 2)}{2} r^2 \Rightarrow$ area seg(2) = 2 × [area seg(1)] QED.

Area of lune = (area small circle)/2 – area seg(2) = $\pi r^2 - \dfrac{(\pi - 2)}{2} r^2 =$ $(\pi + 1)r^2$

12.2: Figure 12.4 from OB clay tablet BM 15285 presents drawing-compass constructions and problems that are hardly less sophisticated than the problem composed by Hippocrates almost 1,500 years later. Figure FQ 12.2 extracts one of the problems from BM 15285. Calculate the area of the shaded part of the drawing.

Area = $(4 - \pi)r^2$

APPENDIX B
DERIVATION OF
EQUATION (11.2)

W aerden (see Waerden reference to chapter 4, p. 77) discusses a cuneiform clay tablet, AO 6484, previously translated by Neugebauer, that contains a calculation of a sum of squares. He knew that it had been composed in the Seleucid era (ca. 200 BCE), but he thought that Babylonian mathematics had sufficiently stagnated after the OB era for this tablet to reflect OB understanding. I have not been able to derive this equation using only documented OB techniques, but I have been able to derive it by using a reasonable (?) extension of a known OB visualization shown in figure 12.2, which also shows how this visualization can be digitalized. The illustrated digitalization shows that $4^2 = 1 + 3 + 5 + 7$, which can be generalized as: the square of any number is the sum gnomons that define the odd numbers, previously given as equation (12.2):

$$n^2 = 1 + 3 + 5 + \ldots (2n - 1) \tag{B.1}$$

Numbers derived from digitalizing geometric figures are known in modern math jargon as figurate numbers. Thus, from figure 12.2, 4^2 is a square number. Derivation of equation (B.1) is generally credited to Pythagoras (580–500 BCE) well before the composition of AO 6468, but so is the Pythagorean theorem, which is now known to have been discovered by the OB period. Thus, it is possible that OB mathematicians knew equation (B.1) or a visualization of it, but there is no OB documentation for it.

The number of blocks in the model pyramid of figure 11.4 can be generalized as:

$$N(\text{pyramid}) = \sum_{1}^{N} n^2 = 1 + 4 + \ldots + n^2 + \ldots + (N-1)^2 + N^2 \tag{B.2}$$

Using equation (B.1), each squared term on the right-hand side of equation (B.2) can be transformed into an arithmetic series of odd numbers:

227

$$N^2 \quad = \quad 1 + 3 + \ldots + (2n - 1) + \ldots + (2N - 3) + (2N - 1)$$

$$(N - 1)^2 = \quad 1 + 3 + \ldots + (2n - 1) + \ldots + (2N - 3)$$

$$\begin{array}{c} \cdot \\ \cdot \\ \cdot \end{array}$$

$$2^2 \quad = \quad 1 + 3$$

$$1^2 \quad = \quad 1$$

Summing all of the terms in this array we obtain:

$$N\text{(pyramid)} = \sum_{1}^{N} n^2 = N + 3(N - 1) + \ldots + (2n - 1)(N + 1 - n) + \ldots + (2N - 3)2 + (2N - 1)$$

$$N\text{(pyramid)} = \sum_{1}^{N} n^2 = \sum_{1}^{N}(2n-1)(N+1-n) = (2N + 3) \sum_{1}^{N} n \quad N(N + 1) - 2 \sum_{1}^{N} n^2$$

The sum of an arithmetic series, which was known in the OB period, is given by equation (B.3)

$$\sum_{1}^{N} n = N(N + 1)/2 \tag{B.3}$$

Employing it in the preceding equation, we obtain:

$$3N\text{(pyramid)} = 3 \sum_{1}^{N} n^2 = (2N + 3) \, N(N + 1)/2 - N(N + 1) = N(N + 1)(2N + 1)/2$$

$$N\text{(pyramid)} = \sum_{1}^{N} n^2 = N(N + 1)(2N + 1)/6 \tag{B.5}$$

Equation (B.5) was previously given as equation (11.2).

Again employing equation (B.3), equation (B.5) can be rewritten in the form that is given in AO 6484:

$$N\text{(pyramid)} = \sum_{1}^{N} n^2 = (2N/3 + 1/3) \sum_{1}^{N} n \tag{B.6}$$

I could not have derived these results without using symbolic algebra, but once having learned this technique it is hard for me, and probably for anybody, to avoid thinking is such terms. Could OB mathematicians have managed such a complicated calculation, or something similar to it, without symbolic algebra? I doubt it. In fact, AO 6484 is not quite a general derivation but rather is only for $N = 10$, yet it appears to be in a correct generalized form of equation (B.4). Even if it were only calculated in Seleucid times, it is still a surprising derivation without symbolic algebra.

APPENDIX C
PROOF $\sqrt{2}$ IS AN
IRRATIONAL NUMBER

To prove that for the right triangle (a, a, c), $c/a = \sqrt{2} \neq p/q$, where p and q are integers, Hippasos used a method called *reductio ad absurdum*:

1. Assume that $c/a = p/q$ is a *rational, reduced fraction*, where p and q are integers with no common factors. Show that this leads to an absurdity.

2. With this assumption that $c/a = p/q$, $(c/a)^2 = (p/q)^2$.

3. For a triangle (a, a, c), the Pythagorean theorem yields $c^2 = 2a^2$ and therefore $p^2 = 2q^2$, and p^2 must be even because 2 is one of its factors.

4. Hence p must also be even and can be written as $p = 2m$, where m is some integer (even number $\times$ even number = even number; odd number $\times$ odd number = odd number).

5. Hence $p^2 = 4m^2 = 2q^2$ so that $q^2 = 2m^2$ and therefore q^2 and q must also be even.

6. In **Step 4** we proved that p must be even, but in **Step 1** we assumed that p/q is a *reduced fraction* and therefore **q must be odd** in order not to have a factor of 2 in common with p, but in **Step 5** we proved that **q must be even.** The assumption that c/a is a *rational fraction* must be wrong and therefore $\sqrt{2}$ is an *irrational* number.

NOTES

I preferentially refer to recent books that are readily available from most booksellers or local libraries rather than original sources that may be available only to scholars at major academic institutions. Unjustly and unfortunately, this system does not always name the author who should be credited, but the reference will provide such identities. Greek mathematicians who are briefly referred to, primarily to establish dates, are not referenced. Wikipedia is a good introduction to biographical data if you are interested in knowing more about them.

Comments about and specific pages for a reference are given here with the publication identified by an abbreviated format. The complete definition of each publication cited is listed in references in alphabetic order by author for each chapter.

CHAPTER 1. NUMBER SYSTEM BASICS

1. This chapter is an abbreviated version of content from *How Mathematics Happened* and references to sources for this chapter can be found there.

CHAPTER 2. EGYPTIAN NUMBERS AND ARITHMETIC

1. This chapter is an abbreviated version of content from *How Mathematics Happened* and references to sources for this chapter can be found there.

CHAPTER 3. BABYLONIAN NUMBERS AND ARITHMETIC

1. This chapter is an abbreviated version of content from *How Mathematics Happened* and I give references to sources only if they had not previously been given there.

2. Neugebauer and Sachs, *Mathematical Cuneiform Texts*, p. 13. Despite the fact that this is a book intended for translators and scholars of cuneiform mathematics, it is a very readable book. It summarizes in English much of Neugebauer's 1930s work that was previously published in German and also includes newer studies of cuneiform texts in UK and US collections. The cognoscenti refer to it as MCT.

3. O'Connor and Robertson. I have tried, without success, to solicit from them the justifications for their proposal. I think their conjecture is reasonable, but my reasons may or may not coincide with theirs.

I have found no print reference and so I have no choice but to list an Internet source. I have no way of knowing when the results I quote were posted or whether the quote will not be deleted by some future editing of this Web page.

4. Hoyrup, *Lengths-Widths-Surfaces*. This is a book of encyclopedic proportions, but it is primarily intended for translators and scholars of cuneiform mathematics. I thank Jens Hoyrup for correspondence that has been invaluable to me in understanding OB mathematics. But it is clear that there are points of interpretation about which we differ; disagreement implies no lack of respect.

CHAPTER 4. OLD BABYLONIAN "QUADRATIC ALGEBRA" PROBLEM TEXTS

1. Neugebauer, *Exact Sciences in Antiquity*, p. 57ff.

2. Hoyrup, *Lengths-Widths-Surfaces*, p. 104.

3. Neugebauer and Sachs, *Mathematical Cuneiform Texts*, p. 129ff.

4. B. L. van der Waerden (1903–1996) attended lectures by Neugebauer at Gottingen University, Germany, in 1927. A. Aaboe (1922–2007) began his graduate studies with Neugebauer at Brown University, United States, in 1955.

5. Waerden, *Science Awakening 1*, p. 125. Waerden's discussions of cuneiform texts are generally easier to understand than those of Neugebauer. He also gives English translations of some publications that had previously only appeared in German.

6. Neugebauer, *Exact Sciences in Antiquity*, p. 61.

7. Waerden, *Science Awakening 1*, p. 72: "Perhaps they derived them by use of diagrams, such as found in Euclid." Waerden notes that this edition repeats material from the edition printed in 1961. My guess is that the contradiction between this and the previous Waerden quotation in this chapter is just a result of inadequate editing-out of superseded opinion from previous editions.

8. Hoyrup, *Lengths-Widths-Surfaces.*

9. In an e-mail to me, October 23, 2007, Hoyrup wrote that in 1984 he sent a preprint of his cut-and-paste paper to Neugebauer. Their mutual friend Asgar Aaboe conveyed Neugebauer's opinion that the new translation was imputing differences in meaning to synonyms.

10. Robson's e-mail to me, November 13, 2007.

11. Waerden, *Science Awakening 1*, p. 63.

12. Hoyrup, *Lengths-Widths-Surfaces*, p. 162ff.

13. Although Neugebauer died in 1995, after the 1930s he concentrated on ancient astronomy.

14. Hoyrup, *Lengths-Widths-Surfaces*, p. 257ff.

15. Maor, *Pythagorean Theorem*, p. 61. I thank Eli Maor for his helpful correspondence.

16. Waerden, *Science Awakening 1*, p. 72ff.

17. Hoyrup, *Lengths-Widths-Surfaces*, p. 234ff.

CHAPTER 5. PYTHAGOREAN TRIPLES

1. Hoyrup, *Lengths-Widths-Surfaces*, p. 275.

2. Friberg, *Unexpected Links Egypt*, p. 81.

3. Friberg, *Unexpected Links Egypt*, p. 174. Friberg's interpretation of problem text VAT 7531 #1 yields both a (3, 4, 5) and a (7, 24, 25) triple. However, the occurrence of (7, 24, 25) does not appear to have been intentional and is just a coincidental consequence of how a triangle was sectioned. It is doubtful that the scribe who composed the problem was aware of its presence.

4. Neugebauer and Sachs, *Mathematical Cuneiform Texts*, p. 38ff.

5. Euclid, *Elements*, bks. I–II, p. 357. This solution has been attributed to Pythagoras and I shall discuss this derivation in chapter 12.

6. Joseph, *Crest of the Peacock*, p. 230.

7. Neugebauer and Sachs, *Mathematical Cuneiform Texts*, p. 38ff.

8. In *How Mathematics Happened*, I found the entry (45, 60, 75) perplexing

because it could not be easily accounted for by the *pq* method. I have learned something since then and I think I now adequately account for this.

9. In *How Mathematics Happened*, p. 215ff., I discuss the enigma of the purpose of Plimpton 322.

10. Waerden, *Science Awakening 1*, p. 78ff.

11. Robson, "Sherlock Holmes," p. 183ff.

CHAPTER 6. SQUARE ROOT CALCULATIONS

1. Neugebauer, *Exact Sciences in Antiquity*, p. 47.

2. Neugebauer and Sachs, *Mathematical Cuneiform Texts*, p. 42.

3. Hoyrup, *Lengths-Widths-Surfaces*, p. 268ff.

4. Heath, *Archimedes*, p. LXXVIIIff.

5. Heath, *Archimedes*, p. LXXIX.

6. Heath, *Archimedes*, p. 91ff.

7. Heath, *Archimedes*, p. LXXXff.

8. Heath, *Archimedes*, p. LXXVII.

9. Rudman, *How Mathematics Happened*, p. 292ff.

10. Heath, *Archimedes*, p. LXXIVff.

CHAPTER 7. PI (π)

1. Gillings, *Mathematics of Pharaohs*, p. 139ff.

2. Gillings, *Mathematics of Pharaohs*, p. 194ff.

3. Neugebauer and Sachs, *Mathematical Cuneiform Texts*, p. 44.

4. Neugebauer, *Exact Sciences in Antiquity*, p. 47

5. Heath, *Archimedes*, p. 93

6. Maor, *Pythagorean Theorem*, p. 52.

CHAPTER 8. SIMILAR TRIANGLES (PROPORTIONALITY)

1. Friberg, *Unexpected Links Egypt*, p. 45ff.

2. Neugebauer and Sachs, *Mathematical Cuneiform Texts*, pp. 48–49.
3. Hoyrup, *Lengths-Widths-Surfaces*, p. 231ff.

CHAPTER 9. SEQUENCES AND SERIES

1. Neugebauer and Sachs, *Mathematical Cuneiform Texts*, p. 52ff.
2. Gillings, *Mathematics of Pharaohs*, pp. 173–75.
3. Gillings, *Mathematics of Pharaohs*, p. 170ff.
4. Friberg, *Unexpected Links Egypt*, p. 37.
5. Friberg, *Unexpected Links Egypt*, p. 5.
6. Rudman, *How Mathematics Happened*, p. 161.

CHAPTER 10. OLD BABYLONIAN ALGEBRA

1. Waerden, *Science Awakening 1*, p. 66. The original translation by Neugebauer was in German. Waerden gives numerical data as decimally transcribed sexagesimal numbers. I have converted all numerical data to base-10 and I have also restated the problem to avoid the complications of units of measurement that are irrelevant to us here.

CHAPTER 11. PYRAMID VOLUME

1. Gillings, *Mathematics of Pharaohs*, p. 146ff.
2. Neugebauer and Sachs, *Mathematical Cuneiform Texts*, p. 63ff.
3. Rudman, *How Mathematics Happened*, p. 175ff.
4. Gillings, *Mathematics of Pharaohs*, p. 189ff.
5. Neugebauer and Sachs, *Mathematical Cuneiform Texts*, p. 94ff.
6. Gillings, *Mathematics of Pharaohs*, p. 188.
7. Waerden, *Science Awakening 1*, p. 34ff.
8. Friberg, *Unexpected Links Egypt*, p. 244.

CHAPTER 12. FROM OLD BABYLONIAN SCRIBE TO LATE BABYLONIAN SCRIBE TO PYTHAGORAS TO PLATO

1. Friberg, *Unexpected Links Egypt.*
2. Heidel, *Babylonian Genesis.*
3. Euclid, *Elements*, bks. I–II, p. 356ff.
4. Euclid, *Elements*, bks. I–II, p. 351.
5. Friberg, *Amazing Traces Greek*, p. 127ff.
6. A translation of *Timaeus* is available on www.gutenberg.com.
7. Herodotus, *Histories*, bks. I-53ff.
8. Mlodinow, *Euclid's Window.*
9. Mandelbrot, *Fractal Geometry.*

CHAPTER 13. EUCLID

1. Euclid, *Elements*, bks. I–II, p. 1ff.
2. Hoyrup, *Lengths-Widths-Surfaces*, p. 98.

REFERENCES

CHAPTER 1. NUMBER SYSTEM BASICS

Rudman, P. S. *How Mathematics Happened: The First 50,000 Years*. Amherst, NY: Prometheus Books, 2006.

CHAPTER 2. EGYPTIAN NUMBERS AND ARITHMETIC

Rudman, P. S. *How Mathematics Happened: The First 50,000 Years*. Amherst, NY: Prometheus Books, 2006.

CHAPTER 3. BABYLONIAN NUMBERS AND ARITHMETIC

Hoyrup, J. *Lengths-Widths-Surfaces: A Portrait of Old Babylonian Algebra and Its Kin*. New York: Springer, 2002.

Neugebauer, O., and A. Sachs. *Mathematical Cuneiform Texts*. New Haven, CT: American Oriental Society, 1945.

O'Connor, J. J., and E. F. Robertson. "Heron of Alexandria." http://www-history .mcs.st-andrews.ac.uk/HistTopics/Babylonian_mathematics.html (accessed September 22, 2009).

Rudman, P. S. *How Mathematics Happened: The First 50,000 Years*. Amherst, NY: Prometheus Books, 2006.

CHAPTER 4. OLD BABYLONIAN "QUADRATIC ALGEBRA" PROBLEM TEXTS

Aaboe, A. *Episodes from the Early History of Mathematics*. New York: Mathematical Association of America, 1997.

Hoyrup, J. *Lengths-Widths-Surfaces: A Portrait of Old Babylonian Algebra and Its Kin*. New York: Springer, 2002.

Maor, E. *The Pythagorean Theorem*. Princeton, NJ: Princeton University Press, 2007.

Neugebauer, O. *The Exact Sciences in Antiquity*. New York: Dover Publications, 1969.

Neugebauer, O., and A. Sachs. *Mathematical Cuneiform Texts*. New Haven, CT: American Oriental Society, 1945.

Waerden, B. L. van der. *Science Awakening 1*. New York: Scholar's Bookshelf, 2005.

CHAPTER 5. PYTHAGOREAN TRIPLES

Euclid. *The Elements*. There are many translations, but all references in this book are to Heath, T. L. *The Thirteen Books of Euclid's Elements*. 1908; New York: Dover, 1956.

Friberg, J. *Unexpected Links between Egyptian and Babylonian Mathematics*. Singapore: World Scientific Publishing, 2005.

Hoyrup, J. *Lengths-Widths-Surfaces: A Portrait of Old Babylonian Algebra and Its Kin*. New York: Springer, 2002.

Joseph, G. G. *The Crest of the Peacock*. Princeton, NJ: Princeton University Press, 2000.

Neugebauer, O., and A. Sachs. *Mathematical Cuneiform Texts*. New Haven, CT: American Oriental Society, 1945.

Robson, E. "Neither Sherlock Holmes nor Babylon: A Reassessment of Plimpton 322." *Historica Mathematica* 28 (2001):183ff.

Rudman, P. S. *How Mathematics Happened: The First 50,000 Years*. Amherst, NY: Prometheus Books, 2006.

Waerden, B. L. van der. *Science Awakening 1*. New York: Scholar's Bookshelf, 2005.

CHAPTER 6. SQUARE ROOT CALCULATIONS

Heath, T. L. *The Works of Archimedes*. 1897; New York: Dover, 2002.

Hoyrup, J. *Lengths-Widths-Surfaces: A Portrait of Old Babylonian Algebra and Its Kin*. New York: Springer, 2002.

Neugebauer, O. *The Exact Sciences in Antiquity*. New York: Dover, 1969.

Neugebauer, O., and A. Sachs. *Mathematical Cuneiform Texts*. New Haven, CT: American Oriental Society, 1945.

Rudman, P. S. *How Mathematics Happened: The First 50,000 Years*. Amherst, NY: Prometheus Books, 2006.

CHAPTER 7. PI (π)

Gillings, R. J. *Mathematics in the Time of the Pharaohs*. New York: Dover Publications, 1962.

Heath, T. L. *The Works of Archimedes*. 1897; New York: Dover, 2002.

Maor, E. *The Pythagorean Theorem*. Princeton, NJ: Princeton University Press, 2007.

Neugebauer, O. *The Exact Sciences in Antiquity*. New York: Dover Publications, 1969.

Neugebauer, O., and A. Sachs. *Mathematical Cuneiform Texts*. New Haven, CT: American Oriental Society, 1945.

CHAPTER 8. SIMILAR TRIANGLES (PROPORTIONALITY)

Friberg, J. *Unexpected Links between Egyptian and Babylonian Mathematics*. Singapore: World Scientific Publishing, 2005.

Hoyrup, J. *Lengths-Widths-Surfaces: A Portrait of Old Babylonian Algebra and Its Kin*. New York: Springer, 2002.

Neugebauer, O., and A. Sachs. *Mathematical Cuneiform Texts*. New Haven, CT: American Oriental Society, 1945.

CHAPTER 9. SEQUENCES AND SERIES

Friberg, J. *Unexpected Links between Egyptian and Babylonian Mathematics*. Singapore: World Scientific Publishing, 2005.

Gillings, R. J. *Mathematics in the Time of the Pharaohs*. New York: Dover, 1982.

Neugebauer, O., and A. Sachs. *Mathematical Cuneiform Texts*. New Haven, CT: American Oriental Society, 1945.

Rudman, P. S. *How Mathematics Happened: The First 50,000 Years*. Amherst, NY: Prometheus Books, 2006.

CHAPTER 10. OLD BABYLONIAN ALGEBRA: SIMULTANEOUS LINEAR EQUATIONS

Waerden, B. L. van der. *Science Awakening 1*. New York: Scholar's Bookshelf, 2005.

CHAPTER 11. PYRAMID VOLUME

Friberg, J. *Unexpected Links between Egyptian and Babylonian Mathematics*. Singapore: World Scientific Publishing, 2005.

Gillings, R. J. *Mathematics in the Time of the Pharaohs*. New York: Dover Publications, 1982.

Neugebauer, O., and A. Sachs. *Mathematical Cuneiform Texts*. New Haven, CT: American Oriental Society, 1945.

Rudman, P. S. *How Mathematics Happened: The First 50,000 Years*. Amherst, NY: Prometheus Books, 2006.

Waerden, B. L. van der. *Science Awakening 1*. New York: Scholar's Bookshelf, 2005.

CHAPTER 12. FROM OLD BABYLONIAN SCRIBE TO LATE BABYLONIAN SCRIBE TO PYTHAGORAS TO PLATO

Euclid. *The Elements*. Translated by T. L. Heath in *The Thirteen Books of Euclid's Elements*. 1908; New York: Dover, 1956.

Friberg, J. *Amazing Traces of a Babylonian Origin in Greek Mathematics*. Singapore: World Scientific Publishing, 2007.

———. *Unexpected Links between Egyptian and Babylonian Mathematics*. Singapore: World Scientific Publishing, 2005.

Heidel, A. *The Babylonian Genesis.* Chicago: University of Chicago Press, 1951.

Herodotus. *The Histories.* There are many fine translations; I quote from a modern language version: Waterfield, R., trans. *The Histories.* New York: Oxford University Press, 1998.

Mandelbrot, B. 1983. *The Fractal Geometry of Nature.* New York: W. H. Freeman, 1983.

Mlodinow, L. *Euclid's Window: The Story of Geometry from Parallel Lines to Hyperspace.* New York: Free Press, 2001.

CHAPTER 13. EUCLID

Euclid. *The Elements.* Translated by T. L. Heath in *The Thirteen Books of Euclid's Elements.* 1908; New York: Dover, 1956.

Hoyrup, J. *Lengths-Widths-Surfaces: A Portrait of Old Babylonian Algebra and Its Kin.* New York: Springer, 2002.

INDEX